Ben Stacy Jerrik (Ed.)

Hepatoporus

Ben Stacy Jerrik (Ed.)

Hepatoporus

Superfamily, Crabs, Families, Atelecy, Clidae, Cancridae

Part Press

Imprint

Publisher:
Part Press is a trademark of
International Book Market Service Ltd., 17 Rue Meldrum, Beau Bassin, 1713-01 Mauritius
Email: info@bookmarketservice.com
Website: www.bookmarketservice.com

Published in 2013

Printed in: U.S.A., U.K., Germany. This book was not produced in Mauritius.

ISBN: 978-620-1-56346-9

Contents

Articles

References

Hepatoporus

Hepatoporus	
Scientific classification	
Kingdom:	Animalia
Phylum:	Arthropoda
Subphylum:	Crustacea
Class:	Malacostraca
Order:	Decapoda
Infraorder:	Brachyura
Family:	Xanthidae
Genus:	***Hepatoporus*** Serène, 1984

Hepatoporus is a genus of crabs in the family Xanthidae, containing the following species:[]

- *Hepatoporus asper* Davie & Turner, 1994
- *Hepatoporus distinctus* (Takeda & Nagai, 1986)
- *Hepatoporus guinotae* (Zarenkov, 1971)
- *Hepatoporus orientalis* (Sakai, 1935)
- *Hepatoporus pumex* Mendoza & Ng, 2008

References

Crab

Crabs	

Liocarcinus vernalis	
Scientific classification	
Kingdom:	Animalia
Phylum:	Arthropoda
Subphylum:	Crustacea
Class:	Malacostraca
Order:	Decapoda
Suborder:	Pleocyemata
Infraorder:	**Brachyura** Linnaeus, 1758
Sections & subsections []	

- Dromiacea
- Raninoida
- Cyclodorippoida
- Eubrachyura
  - Heterotremata
  - Thoracotremata

Crabs are decapod crustaceans of the infraorder **Brachyura**, which typically have a very short projecting "tail" (abdomen) (Greek: βραχύς / *brachys* = short,[1] οὐρά / *oura* = tail[2]), usually entirely hidden under the thorax. They live in all the world's oceans, in fresh water, and on land, are generally covered with a thick exoskeleton and have a single pair of claws. Many other animals with similar names – such as hermit crabs, king crabs, porcelain crabs, horseshoe crabs and crab lice – are not true crabs.

Evolution

Gecarcinus quadratus, a land crab from Central and South America

Crabs are generally covered with a thick exoskeleton, composed primarily of calcium carbonate,[3] and armed with a single pair of chelae (claws). Crabs are found in all of the world's oceans, while many crabs live in fresh water and on land, particularly in tropical regions. Crabs vary in size from the pea crab, a few millimetres wide, to the Japanese spider crab, with a leg span of up to 4 metres (ft).[4]

About 850 species of crab are freshwater, terrestrial or semi-terrestrial species;[5] they are found throughout the world's tropical and semi-tropical regions. They were previously thought to be a monophyletic group, but are now believed to represent at least two distinct lineages, one in the Old World and one in the New World.[6]

The earliest unambiguous crab fossils date from the Jurassic,[7] although Carboniferous *Imocaris*, known only from its carapace, may be a primitive crab.[8] The radiation of crabs in the Cretaceous and afterward may be linked either to the break-up of Gondwana or to the concurrent radiation of bony fish, crabs' main predators.[9]

Sexual dimorphism

Crabs often show marked sexual dimorphism. Males often have larger claws,[10] a tendency which is particularly pronounced in the fiddler crabs of the genus *Uca* (Ocypodidae). In fiddler crabs, males have one claw which is greatly enlarged and which is used for communication, particularly for attracting a mate.[11] Another conspicuous difference is the form of the pleon (abdomen); in most male crabs, this is narrow and triangular in form, while females have a broader, rounded abdomen.[12] This is due to the fact that female crabs brood fertilised eggs on their pleopods.

The underside of a male (top) and a female (bottom) individual of *Pachygrapsus marmoratus*, showing the difference in shape of the abdomen

Reproduction and life cycle

Crabs attract a mate through chemical (pheromones), visual, acoustic or vibratory means. Pheromones are used by most fully aquatic crabs, while terrestrial and semi-terrestrial crabs often use visual signals, such as fiddler crab males waving their large claw to attract females. The vast number of brachyuran crabs have internal fertilisation and mate belly-to-belly. For many aquatic species, mating takes place just after the female has moulted and is still soft. Females can store the sperm for a long time before using it to fertilise their eggs. When fertilisation has taken place, the eggs are released onto the female's abdomen, below the tail flap, secured with a sticky material. In this location they are protected during embryonic development. Females carrying eggs are called "berried" since the eggs resemble round berries.

When development is complete, the female releases the newly-hatched larvae into the water, where they are part of the plankton. The release is often timed with the tides. The free-swimming tiny zoea larvae can float and take advantage of water currents. They have a spine, which probably reduces the rate of predation by larger animals. The zoea of most species must find food, but some crabs provide enough yolk in the eggs that the larval stages can continue to live off the yolk.

Each species has a particular number of zoeal stages, separated by moults, before they change into a megalopa stage, which resembles an adult crab, except for having the abdomen (tail) sticking out behind. After one more moult, the crab is a juvenile, living on the bottom rather than floating in the water. This last moult, from megalopa to juvenile is critical, and it must take place in a habitat that is suitable for the juvenile to survive.[] :63–77

Female crab *Xantho poressa* at spawning time in the Black Sea, carrying eggs under her abdomen.

Most species of terrestrial crabs must migrate down to the ocean to release their larvae; in some cases this entails very extensive migrations. After living for a short time as larvae in the ocean, the juveniles must do this migration in reverse. In many tropical areas with land crabs, these migrations often result in considerable roadkill of migrating crabs.[] :113–114

Once crabs have become juveniles they will still have to keep moulting many more times to become adults. They are covered with a hard shell, which would otherwise prevent growth. The moult cycle is coordinated by hormones. When preparing for moult, the old shell is softened and partly eroded away, while the rudimentary beginnings of a new shell form under it. At the time of moulting, the crab takes in a lot of water to expand and crack open the old shell at a line of weakness along the back edge of the carapace. The crab must then extract all of itself – including its legs, mouthparts, eyestalks, and even the lining of the front and back of the digestive tract – from the old shell. This is a difficult process that takes many hours, and if a crab gets stuck it will die. After freeing itself from the old shell (now called an exuvia) the crab is extremely soft and hides until its new shell has hardened. While the new shell is still soft, the crab can expand it to make room for future growth.[] :78–79

Behaviour

Carpilius convexus consuming *Heterocentrotus trigonarius* in Hawaii

Crabs typically walk sideways[] (a behaviour which gives us the word *crabwise*). This is because of the articulation of the legs which makes a sidelong gait more efficient.[] However, some crabs walk forwards or backwards, including raninids,[13] *Libinia emarginata*[14] and *Mictyris platycheles*.[] Some crabs, notably the Portunidae and Matutidae, are also capable of swimming,[] the Portunidae especially so as their last pair of walking legs are flattened into swimming paddles.[] :96

Crabs are mostly active animals with complex behaviour patterns. They can communicate by drumming or waving their pincers. Crabs tend to be aggressive towards one another and males often fight to gain access to females.[15] On rocky seashores, where nearly all caves and crevices are occupied, crabs may also fight over hiding holes.[16] Fiddler crabs (genus *Uca*) dig burrows in sand or mud, which they use for resting, hiding, mating and to defend against intruders.[] :28–29, 99

Crabs are omnivores, feeding primarily on algae,[17] and taking any other food, including molluscs, worms, other crustaceans, fungi, bacteria and detritus, depending on their availability and the crab species. For many crabs, a mixed diet of plant and animal matter results in the fastest growth and greatest fitness.[18] [19] However, some species are more specialised in their diets. Some eat plankton, some eat primarily shellfish like clams and some even catch fish.[] :85

Crabs are known to work together to provide food and protection for their family, and during mating season to find a comfortable spot for the female to release her eggs.[20]

Human consumption

Fisheries

Fishermen sorting velvet crabs at Fionnphort, Scotland

Crabs make up 20% of all marine crustaceans caught, farmed, and consumed worldwide, amounting to 1½ million tonnes annually. One species, *Portunus trituberculatus*, accounts for one fifth of that total. Other commercially important taxa include *Portunus pelagicus*, several species in the genus *Chionoecetes*, the blue crab (*Callinectes sapidus*), *Charybdis spp.*, *Cancer pagurus*, the Dungeness crab (*Metacarcinus magister*) and *Scylla serrata*, each of which yields more than 20,000 tonnes annually.[21]

In some species, crab meat is harvested by manually twisting and pulling off one or both claws and returning the live crab to the water in the belief the crab will survive and regenerate the claws thereby making it a sustainable industry.[22] [23] [24] [25]

Cookery

Further information: Category:Crab dishes

Crab masala from Karnataka, India

Crabs are prepared and eaten as a dish in several different ways all over the world. Some species are eaten whole, including the shell, such as soft-shell crab; with other species just the claws and/or legs are eaten. The latter is particularly common for larger crabs, such as the snow crab. Mostly in East Asian cultures, the roe of the female crab is also eaten, which usually appears orange or yellow in colour in fertile crabs.

The biggest importers, and therefore countries where eating crab is immensely popular, are Japan, France, Spain, Hong Kong, the US, Canada and Portugal.[26]

In some regions spices improve the culinary experience. In Southeast Asia and Indosphere, masala crab and chilli crab are examples of heavily spiced dishes. In the Chesapeake Bay region, blue crab is often eaten with Old Bay Seasoning. Alaskan king crab or snow crab legs are usually simply boiled and served with garlic butter.

For the British dish Cromer crab, the crab meat is extracted and placed inside the hard shell. One American way to prepare crab meat is by extracting it and adding a flour mix, creating a crab cake.

Crabs are also used in bisque, a global dish of French origin.

Pain

Crabs are often boiled alive. In 2005, Norwegian scientists concluded that lobsters cannot feel pain.[27] However, later research suggests that crustaceans are indeed able to feel and remember pain.[28] [29]

Classification

The infraorder Brachyura contains 6,793 species in 93 families,[] as many as the remainder of the Decapoda.[] The evolution of crabs is characterised by an increasingly robust body, and a reduction in the abdomen. Although many other groups have undergone similar processes, carcinisation is most advanced in crabs. The telson is no longer functional in crabs, and the uropods are absent, having probably evolved into small devices for holding the reduced abdomen tight against the sternum.

In most decapods, the gonopores (sexual openings) are found on the legs. However, since crabs use the first two pairs of pleopods (abdominal appendages) for sperm transfer, this arrangement has changed. As the male abdomen evolved into a narrower shape, the gonopores have moved towards the midline, away from the legs, and onto the sternum.[30] A similar change occurred, independently, with the female gonopores. The movement of the female gonopore to the sternum defines the clade Eubrachyura, and the later change in the position of the male gonopore defines the Thoracotremata. It is still a subject of debate whether those crabs where the female, but not male, gonopores are situated on the sternum, form a monophyletic group.[]

Superfamilies

Numbers of extant and extinct (†) species are given in brackets.[] The superfamily Eocarcinoidea, containing *Eocarcinus* and *Platykotta*, was formerly thought to contain the oldest crabs; it is now considered part of the Anomura.[]

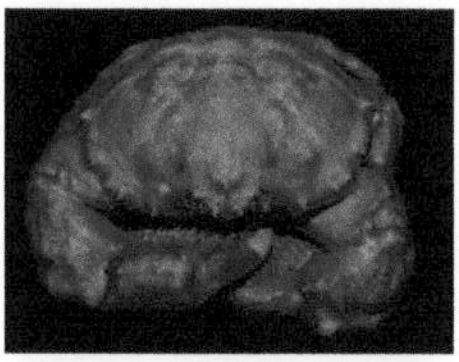

Dromia personata (Dromiacea: Dromiidae)

Ranina ranina (Raninoida: Raninidae)

Corystes cassivelaunus (Heterotremata: Corystidae)

Ocypode quadrata (Thoracotremata: Ocypodidae)

- **Section Dromiacea**
  - Dakoticancroidea (6†)
  - Dromioidea (147, 85†)
  - Glaessneropsoidea (45†)

- Homolodromioidea (24, 107†)
- Homoloidea (73, 49†)
- **Section Raninoida** (46, 196†)
- **Section Cyclodorippoida** (99, 27†)
- **Section Eubrachyura**
  - **Subsection Heterotremata**
  - Aethroidea (37, 44†)
  - Bellioidea (7)
  - Bythograeoidea (14)
  - Calappoidea (101, 71†)
  - Cancroidea (57, 81†)
  - Carpilioidea (4, 104†)
  - Cheiragonoidea (3, 13†)
  - Corystoidea (10, 5†)
  - Componocancroidea (1†)
  - Dairoidea (4, 8†)
  - Dorippoidea (101, 73†)
  - Eriphioidea (67, 14†)
  - Gecarcinucoidea (349)
  - Goneplacoidea (182, 94†)
  - Hexapodoidea (21, 25†)
  - Leucosioidea (488, 113†)
  - Majoidea (980, 89†)
  - Orithyioidea (1)
  - Palicoidea (63, 6†)
  - Parthenopoidea (144, 36†)
  - Pilumnoidea (405, 47†)
  - Portunoidea (455, 200†)
  - Potamoidea (662, 8†)
  - Pseudothelphusoidea (276)
  - Pseudozioidea (22, 6†)
  - Retroplumoidea (10, 27†)
  - Trapezioidea (58, 10†)
  - Trichodactyloidea (50)
  - Xanthoidea (736, 134†)
  - **Subsection Thoracotremata**
  - Cryptochiroidea (46)
  - Grapsoidea (493, 28†)
  - Ocypodoidea (304, 14†)
  - Pinnotheroidea (304, 13†)

Cultural influences

Both the constellation Cancer and the astrological sign Cancer are named after the crab, and depicted as a crab. William Parsons, 3rd Earl of Rosse drew the Crab Nebula in 1848 and noticed its similarity to the animal; the Crab pulsar lies at the centre of the nebula.[31]

The Moche people of ancient Peru worshipped nature, especially the sea,[32] and often depicted crabs in their art.[33] In Greek mythology, Karkinos was a crab that came to the aid of the Lernaean Hydra as it battled Heracles.

Crabs are also prevalent in pop culture including films and TV programs. Sebastian in *The Little Mermaid* is a popular character who sings the famous song "Under the Sea". Others are not quite so likeable. Mr. Krabs, the crab in *SpongeBob SquarePants*, is money-obsessed and not very likeable. Gigantic monster crabs are featured in some Hollywood movies such as *Attack of the Crab Monsters,* a "camp" horror movie of the 1950s. Crabs are the subject of annual festivals in many coastal states, particularly because they are the featured food there. Images of crabs are also popular in jewellery and t-shirts.[] :145, 211–213

References

[26] Guide to crabs - types of crab and crabmeat plus information on how to choose, buy and store your live crabs. (http://www.helpwithcooking.com/seafood-shellfish/guide-to-crabs.html)

Xanthidae

Xanthidae	

Atergatis subdentatus	
Scientific classification	
Kingdom:	Animalia
Phylum:	Arthropoda
Subphylum:	Crustacea
Class:	Malacostraca
Order:	Decapoda
Infraorder:	Brachyura
Superfamily:	Xanthoidea
Family:	**Xanthidae** Macleay, 1838 []
Subfamilies	

- Actaeinae Alcock, 1898
- Antrocarcininae Ng & D. G. B. Chia, 1994
- Chlorodiellinae Ng & Holthuis, 2007
- Cymoinae Alcock, 1898
- Etisinae Ortmann, 1893
- Euxanthinae Alcock, 1898
- Glyptoxanthinae Mendoza & Guinot, 2011
- Kraussiinae Ng, 1993
- Liomerinae T. Sakai, 1976
- Polydectinae Dana, 1851
- Speocarcininae Števčić, 2005
- Xanthinae MacLeay, 1838
- Zalasiinae Serène, 1968
- Zosiminae Alcock, 1898

Xanthidae is a family of crabs known as **mud crabs**, **pebble crabs** or **rubble crabs**.[] They are often brightly coloured and are poisonous, containing toxins which are not destroyed by cooking and for which no antidote is known.[] The toxins are similar to the tetrodotoxin and saxitoxin produced by puffer fish, and may be produced by bacteria in the genus *Vibrio* living in symbiosis with the crabs, mostly *V. alginolyticus* and *V. parahaemolyticus*.[]

Classification

Many species formerly included in the family Xanthidae have since been moved to new families. Despite this, Xanthidae is still the largest crab family in terms of species richness, with 572 species in 133 genera divided among the thirteen subfamilies:[]

- Actaeinae Alcock, 1898
  - *Actaea* De Haan, 1833
  - *Actaeodes* Dana, 1851
  - *Actaeops* † Portell & Collins, 2004
  - *Allactaea* Williams, 1974
  - *Epiactaea* Serène, 1984
  - *Epiactaeodes* Serène, 1984
  - *Forestia* Guinot, 1976
  - *Gaillardiellus* Guinot, 1976
  - *Heteractaea* Lockington, 1877
  - *Lobiactaea* T. Sakai, 1983
  - *Meractaea* Serène, 1984
  - *Novactaea* Guinot, 1976
  - *Odhneria* T. Sakai, 1983
  - *Paractaea* Guinot, 1969
  - *Paractaeopsis* Serène, 1984
  - *Phlyctenodes* † A. Milne-Edwards, 1862
  - *Platyactaea* Guinot, 1967
  - *Psaumis* Kossmann, 1877
  - *Pseudactaea* Serène, 1962
  - *Pseudoliomera* Odhner, 1925
  - *Pseudophlyctenodes* † Busulini, Tessier & Beschin, 2006
  - *Rata* Davie, 1993
  - *Serenius* Guinot, 1976
- Antrocarcininae Ng & D. G. B. Chia, 1994
  - *Antrocarcinus* Ng & D. G. B. Chia, 1994
  - *Cyrtocarcinus* Ng & D. G. B. Chia, 1994
  - *Glyptocarcinus* Takeda, 1973
- Chlorodiellinae Ng & Holthuis, 2007
  - *Chlorodiella* Rathbun, 1897
  - *Cyclodius* Dana, 1851
  - *Garthiella* Titgen, 1986
  - *Liocarpiloides* Klunzinger, 1913
  - *Pilodius* Dana, 1851
  - *Sulcodius* P. F. Clark & Ng, 1999
  - *Tweedieia* Ward, 1935
  - *Vellodius* Ng & Yang, 1998
- Cymoinae Alcock, 1898
  - *Cymo* De Haan, 1833
- Etisinae Ortmann, 1893
  - *Etisus* H. Milne-Edwards, 1834
  - *Paraetisus* Ward, 1933

- Euxanthinae Alcock, 1898
  - *Alainodaeus* Davie, 1993
  - *Batodaeus* Vázquez-Bader & Gracia, 2004
  - *Carpoporus* Stimpson, 1871
  - *Cranaothus* Ng, 1993
  - *Crosnierius* Serène & Vadon, 1981
  - *Danielea* Ng & P. F. Clark, 2003
  - *Edwardsium* Guinot, 1967
  - *Epistocavea* Davie, 1993
  - *Euxanthus* Dana, 1851
  - *Guinotellus* Serène, 1971
  - *Hepatoporus* Serène, 1984
  - *Hypocolpus* Rathbun, 1897
  - *Ladomedaeus* Števčić, 2005
  - *Lipaesthesius* Rathbun, 1898
  - *Medaeops* Guinot, 1967
  - *Medaeus* Dana, 1851
  - *Miersiella* Guinot, 1967
  - *Monodaeus* Guinot, 1967
  - *Olenothus* Ng, 2002
  - *Palatigum* Davie, 1997
  - *Paramedaeus* Guinot, 1967
  - *Pleurocolpus* Crosnier, 1995
  - *Pseudomedaeus* Guinot, 1968
  - *Rizalthus* Mendoza & Ng, 2008
  - *Visayax* Mendoza & Ng, 2008
- Glyptoxanthinae Mendoza & Guinot, 2011 [1]
  - *Glyptoxanthus* A. Milne-Edwards, 1879
- Kraussiinae Ng, 1993
  - *Garthasia* Ng, 1993
  - *Kraussia* Dana, 1852
  - *Palapedia* Ng, 1993
- Liomerinae T. Sakai, 1976
  - *Actiomera* Ng, Guinot & Davie, 2008
  - *Bruciana* Serène, 1977
  - *Liomera* Dana, 1851
  - *Meriola* Davie, 1993
  - *Neoliomera* Odhner, 1925
  - *Neomeria* † C.-H. Hu & Tao, 1996
  - *Paraliomera* Rathbun, 1930
- Polydectinae Dana, 1851
  - *Lybia* H. Milne-Edwards, 1834
  - *Polydectus* H. Milne-Edwards, 1837
- Speocarcininae Števčić, 2005
  - *Speocarcinus* Stimpson, 1859
- Xanthinae MacLeay, 1838

- *Cataleptodius* Guinot, 1968
- *Coralliope* Guinot, 1967
- *Cycloxanthops* Rathbun, 1897
- *Demania* Laurie, 1906
- *Ectaesthesius* Rathbun, 1898
- *Epixanthops* Serène, 1984
- *Eucratodes* A. Milne-Edwards, 1880
- *Euryxanthops* Garth & Kim, 1983
- *Garthiope* Guinot, 1990
- *Gaudichaudia* Rathbun, 1930
- *Gonopanope* Guinot, 1967
- *Guitonia* Garth & Iliffe, 1992
- *Jacforus* Ng & P. F. Clark, 2003
- *Juxtaxanthias* Ward, 1942
- *Lachnopodus* Stimpson, 1858
- *Leptodius* A. Milne-Edwards, 1863
- *Liagore* De Haan, 1833
- *Linnaeoxanthus* Števčić, 2005
- *Lioxanthodes* Calman, 1909
- *Macromedaeus* Ward, 1942
- *Marratha* Ng & P. F. Clark, 2003
- *Megametope* Filhol, 1886
- *Megamia* † Karasawa, 1993
- *Melybia* Stimpson, 1871
- *Metaxanthops* Serène, 1984
- *Metopoxantho* † De Man, 1904
- *Microcassiope* Guinot, 1967
- *Micropanope* Stimpson, 1871
- *Nanocassiope* Guinot, 1967
- *Nectopanope* Wood-Mason & Alcock, 1891
- *Neolioxantho* Garth & Kim, 1983
- *Neoxanthias* Ward, 1932
- *Neoxanthops* Guinot, 1968
- *Orphnoxanthus* Alcock, 1896
- *Ovatis* Ng & H.-I. Chen, 2004
- *Palaeoxanthops* † Karasawa, 1993
- *Paraxanthias* Odhner, 1925
- *Paraxanthodes* Guinot, 1968
- *Paraxanthus* Lucas, 1844
- *Xanthias* Rathbun, 1897
- *Xantho* Leach, 1814
- *Xanthodius* Stimpson, 1859

- Zalasiinae Serène, 1968
  - *Banareia* A. Milne-Edwards, 1869
  - *Calvactaea* Ward, 1933
  - *Zalasius* Rathbun, 1897
- Zosiminae Alcock, 1898

  - *Atergatis* De Haan, 1833
  - *Atergatopsis* A. Milne-Edwards, 1862
  - *Lophozozymus* A. Milne-Edwards, 1863
  - *Paratergatis* T. Sakai, 1965
  - *Platypodia* Bell, 1835
  - *Platypodiella* Guinot, 1967
  - *Pulcratis* Ng & Huang, 1977
  - *Zosimus* Leach, 1818
  - *Zozymodes* Heller, 1861
- *Incertae sedis*
  - *Haydnella* † Müller, 1984
  - *Nogarolia* † Beschin, Busulini, De Angeli & Tessier, 1994
  - *Sculptoplax* † Müller & Collins, 1991

References

Animal

Animals Temporal range: Ediacaran – Recent	
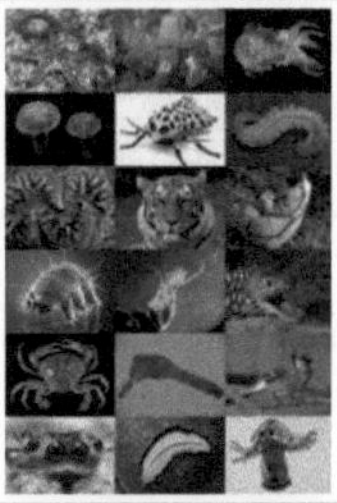	
Scientific classification	
Domain:	Eukaryota
(Unranked)	Opisthokonta
(Unranked)	Holozoa
(Unranked)	Filozoa
Kingdom:	**Animalia** Linnaeus, 1758
Phyla	

- **Subkingdom Parazoa**
  - Porifera
  - Placozoa
- **Subkingdom Eumetazoa**
  - **Radiata (unranked)**
    - Ctenophora
    - Cnidaria
  - **Bilateria (unranked)**
    - Orthonectida
    - Rhombozoa
    - Acoelomorpha
    - Chaetognatha
    - **Superphylum Deuterostomia**
      - Chordata
      - Hemichordata
      - Echinodermata
      - Xenoturbellida
      - Vetulicolia †
    - **Protostomia (unranked)**
      - **Superphylum Ecdysozoa**
        - Kinorhyncha
        - Loricifera
        - Priapulida
        - Nematoda
        - Nematomorpha
        - Lobopodia
        - Onychophora
        - Tardigrada
        - Arthropoda
      - **Superphylum Platyzoa**
        - Platyhelminthes
        - Gastrotricha
        - Rotifera
        - Acanthocephala
        - Gnathostomulida
        - Micrognathozoa
        - Cycliophora
      - **Superphylum Lophotrochozoa**
        - Sipuncula
        - Hyolitha †
        - Nemertea
        - Phoronida
        - Bryozoa
        - Entoprocta
        - Brachiopoda
        - Mollusca
        - Annelida
        - Echiura

Animals are multicellular, eukaryotic organisms of the kingdom **Animalia** or **Metazoa**. Their body plan eventually becomes fixed as they develop, although some undergo a process of metamorphosis later on in their life. Most animals are motile, meaning they can move spontaneously and independently. All animals must ingest other organisms or their products for sustenance (see Heterotroph).

Most known animal phyla appeared in the fossil record as marine species during the Cambrian explosion, about 542 million years ago. Animals are divided into various sub-groups, including birds, mammals, amphibians, reptiles, fish and insects.

Etymology

The word "animal" comes from the Latin word *animalis*, meaning "having breath".[1] In everyday colloquial usage, the word often refers to non-human members of the kingdom Animalia. Sometimes, only closer relatives of humans such as mammals and other vertebrates are meant in colloquial use.[2] The biological definition of the word refers to all members of the kingdom Animalia, encompassing creatures as diverse as sponges, jellyfish, insects, and humans.[]

Characteristics

Animals have several characteristics that set them apart from other living things. Animals are eukaryotic and multicellular,[] which separates them from bacteria and most protists. They are heterotrophic,[] generally digesting food in an internal chamber, which separates them from plants and algae.[3] They are also distinguished from plants, algae, and fungi by lacking rigid cell walls.[] All animals are motile,[] if only at certain life stages. In most animals, embryos pass through a blastula stage,[4] which is a characteristic exclusive to animals.

Structure

With a few exceptions, most notably the sponges (Phylum Porifera) and Placozoa, animals have bodies differentiated into separate tissues. These include muscles, which are able to contract and control locomotion, and nerve tissues, which send and process signals. Typically, there is also an internal digestive chamber, with one or two openings.[] Animals with this sort of organization are called metazoans, or eumetazoans when the former is used for animals in general.[5]

All animals have eukaryotic cells, surrounded by a characteristic extracellular matrix composed of collagen and elastic glycoproteins.[6] This may be calcified to form structures like shells, bones, and spicules.[7] During development, it forms a relatively flexible framework[8] upon which cells can move about and be reorganized, making complex structures possible. In contrast, other multicellular organisms, like plants and fungi, have cells held in place by cell walls, and so develop by progressive growth.[] Also, unique to animal cells are the following intercellular junctions: tight junctions, gap junctions, and desmosomes.[9]

Reproduction and development

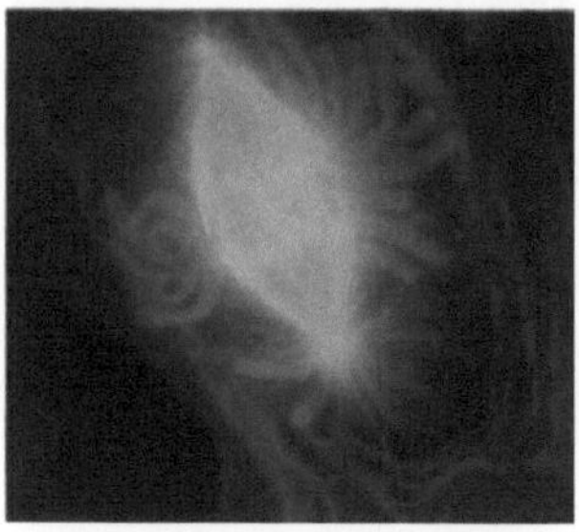

A newt lung cell stained with fluorescent dyes undergoing the early anaphase stage of mitosis

Nearly all animals undergo some form of sexual reproduction.[10] They have a few specialized reproductive cells, which undergo meiosis to produce smaller, motile spermatozoa or larger, non-motile ova.[11] These fuse to form zygotes, which develop into new individuals.[12]

Many animals are also capable of asexual reproduction.[13] This may take place through parthenogenesis, where fertile eggs are produced without mating, budding, or fragmentation.[14]

A zygote initially develops into a hollow sphere, called a blastula,[15] which undergoes rearrangement and differentiation. In sponges, blastula larvae swim to a new location and develop into a new sponge.[16] In most other groups, the blastula undergoes more complicated rearrangement.[17] It first invaginates to form a gastrula with a digestive chamber, and two separate germ layers — an external ectoderm and an internal endoderm.[18] In most cases, a mesoderm also develops between them.[19] These germ layers then differentiate to form tissues and organs.[20]

Food and energy sourcing

All animals are heterotrophs, meaning that they feed directly or indirectly on other living things.[21] They are often further subdivided into groups such as carnivores, herbivores, omnivores, and parasites.[22]

Predation is a biological interaction where a predator (a heterotroph that is hunting) feeds on its prey (the organism that is attacked).[23] Predators may or may not kill their prey prior to feeding on them, but the act of predation always results in the death of the prey.[24] The other main category of consumption is detritivory, the consumption of dead organic matter.[25] It can at times be difficult to separate the two feeding behaviours, for example, where parasitic species prey on a host organism and then lay their eggs on it for their offspring to feed on its decaying corpse. Selective pressures imposed on one another has led to an evolutionary arms race between prey and predator, resulting in various antipredator adaptations.[26]

Most animals indirectly use the energy of sunlight by eating plants or plant-eating animals. Most plants use light to convert inorganic molecules in their environment into carbohydrates, fats, proteins and other biomolecules, characteristically containing reduced carbon in the form of carbon-hydrogen bonds. Starting with carbon dioxide (CO_2) and water (H_2O), photosynthesis converts the energy of sunlight into chemical energy in the form of simple sugars (e.g., glucose), with the release of molecular oxygen. These sugars are then used as the building blocks for plant growth, including the production of other biomolecules.[] When an animal eats plants (or eats other animals which have eaten plants), the reduced carbon compounds in the food become a source of energy and building materials for the animal.[27] They are either used directly to help the animal grow, or broken down, releasing stored solar energy, and giving the animal the energy required for motion.[28] [29]

Animals living close to hydrothermal vents and cold seeps on the ocean floor are not dependent on the energy of sunlight.[30] Instead chemosynthetic archaea and bacteria form the base of the food chain.[31]

Origin and fossil record

Further information: Urmetazoan

Animals are generally considered to have evolved from a flagellated eukaryote.[33] Their closest known living relatives are the choanoflagellates, collared flagellates that have a morphology similar to the choanocytes of certain sponges.[34] Molecular studies place animals in a supergroup called the opisthokonts, which also include the choanoflagellates, fungi and a few small parasitic protists.[35] The name comes from the posterior location of the flagellum in motile cells, such as most animal spermatozoa, whereas other eukaryotes tend to have anterior flagella.[36]

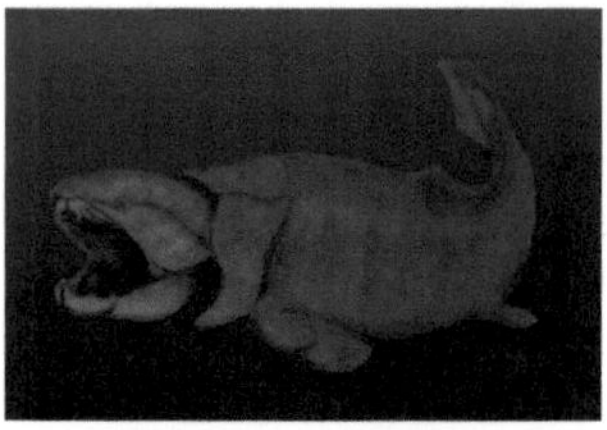

Dunkleosteus was a 10-metre-long (ft) prehistoric fish.[32]

The first fossils that might represent animals appear in the Trezona Formation at Trezona Bore, West Central Flinders, South Australia.[] These fossils are interpreted as being early sponges. They were found in 665-million-year-old rock.[]

The next oldest possible animal fossils are found towards the end of the Precambrian, around 610 million years ago, and are known as the Ediacaran or Vendian biota.[37] These are difficult to relate to later fossils, however. Some may represent precursors of modern phyla, but they may be separate groups, and it is possible they are not really animals at all.[38]

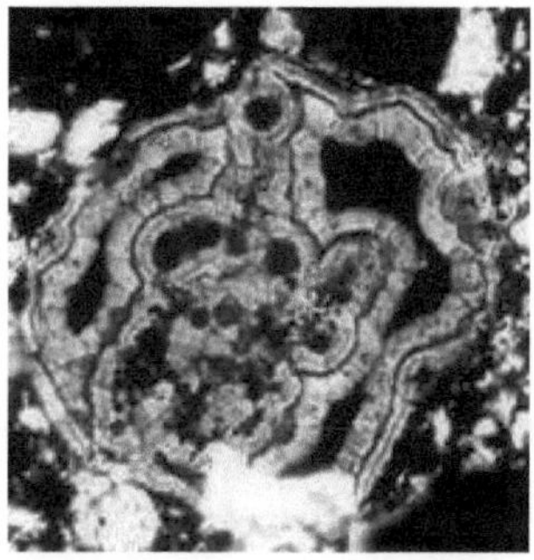

Vernanimalcula guizhouena is a fossil believed by some to represent the earliest known member of the Bilateria.

Aside from them, most known animal phyla make a more or less simultaneous appearance during the Cambrian period, about 542 million years ago.[39] It is still disputed whether this event, called the Cambrian explosion, represents a rapid divergence between different groups or a change in conditions that made fossilization possible.

Some paleontologists suggest that animals appeared much earlier than the Cambrian explosion, possibly as early as 1 billion years ago.[40] Trace fossils such as tracks and burrows found in the Tonian era indicate the presence of triploblastic worms, like metazoans, roughly as large (about 5 mm wide) and complex as earthworms.[] During the beginning of the Tonian period around 1 billion years ago, there was a decrease in Stromatolite diversity, which may indicate the appearance of grazing animals, since stromatolite diversity increased when grazing animals went extinct at the End Permian and End Ordovician extinction events, and decreased shortly after the grazer populations recovered. However the discovery that tracks very similar to these early trace fossils are produced today by the giant single-celled protist *Gromia sphaerica* casts doubt on their interpretation as evidence of early animal evolution.[] []

It has been estimated that 99.9% of animals that have ever existed are extinct.[41]

Groups of animals

Porifera, Radiata and basal Bilateria

Phylogenetic analysis suggests that the Porifera and Ctenophora diverged before a clade that gave rise to the Bilateria, Cnidaria and Placozoa.[] Another study based on the presence/absence of introns suggests that Cnidaria, Porifera and Placozoa may be a sister group of Bilateria and Ctenophora.[42]

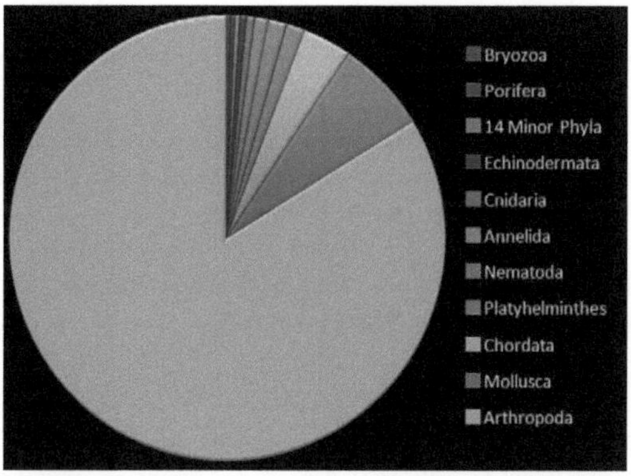

The relative number of species contributed to the total by each phylum of animals.

Orange elephant ear sponge, *Agelas clathrodes*, in foreground. Two corals in the background: a sea fan, *Iciligorgia schrammi*, and a sea rod, *Plexaurella nutans*.

The sponges (Porifera) were long thought to have diverged from other animals early.[43] They lack the complex organization found in most other phyla.[44] Their cells are differentiated, but in most cases not organized into distinct tissues.[45] Sponges typically feed by drawing in water through pores.[46] Archaeocyatha, which have fused skeletons, may represent sponges or a separate phylum.[47] However, a phylogenomic study in 2008 of 150 genes in 29 animals across 21 phyla revealed that it is the Ctenophora or comb jellies which are the basal lineage of animals, at least among those 21 phyla. The authors speculate that sponges—or at least those lines of sponges they investigated—are not so primitive, but may instead be secondarily simplified.[]

Among the other phyla, the Ctenophora and the Cnidaria, which includes sea anemones, corals, and jellyfish, are radially symmetric and have digestive chambers with a single opening, which serves as both the mouth and the anus.[48] Both have distinct tissues, but they are not organized into organs.[49] There are only two main germ layers, the ectoderm and endoderm, with only scattered cells between them. As such, these animals are sometimes called diploblastic.[50] The tiny placozoans are similar, but they do not have a permanent digestive chamber.

The remaining animals form a monophyletic group called the Bilateria. For the most part, they are bilaterally symmetric, and often have a specialized head with feeding and sensory organs. The body is triploblastic, i.e. all three germ layers are well-developed, and tissues form distinct organs. The digestive chamber has two openings, a mouth and an anus, and there is also an internal body cavity called a coelom or pseudocoelom. There are exceptions to each of these characteristics, however — for instance adult echinoderms are radially symmetric, and certain parasitic worms have extremely simplified body structures.

Genetic studies have considerably changed our understanding of the relationships within the Bilateria. Most appear to belong to two major lineages: the deuterostomes and the protostomes, the latter of which includes the Ecdysozoa, Platyzoa, and Lophotrochozoa. In addition, there are a few small groups of bilaterians with relatively similar structure that appear to have diverged before these major groups. These include the Acoelomorpha, Rhombozoa, and Orthonectida. The Myxozoa, single-celled parasites that were originally considered Protozoa, are now believed to have developed from the Medusozoa as well.

Deuterostomes

Superb Fairy-wren, *Malurus cyaneus*

Deuterostomes differ from the other Bilateria, called protostomes, in several ways. In both cases there is a complete digestive tract. However, in protostomes, the first opening of the gut to appear in embryological development (the archenteron) develops into the mouth, with the anus forming secondarily. In deuterostomes the anus forms first, with the mouth developing secondarily.[51] In most protostomes, cells simply fill in the interior of the gastrula to form the mesoderm, called schizocoelous development, but in deuterostomes, it forms through invagination of the endoderm, called enterocoelic pouching.[52] Deuterostome embryos undergo radial cleavage during cell division, while protostomes undergo spiral cleavage.[53]

All this suggests the deuterostomes and protostomes are separate, monophyletic lineages. The main phyla of deuterostomes are the Echinodermata and Chordata.[54] The former are radially symmetric and exclusively marine, such as starfish, sea urchins, and sea cucumbers.[55] The latter are dominated by the vertebrates, animals with backbones.[56] These include fish, amphibians, reptiles, birds, and mammals.[57]

In addition to these, the deuterostomes also include the Hemichordata, or acorn worms.[58] Although they are not especially prominent today, the important fossil graptolites may belong to this group.[59]

The Chaetognatha or arrow worms may also be deuterostomes, but more recent studies suggest protostome affinities.

Ecdysozoa

The Ecdysozoa are protostomes, named after the common trait of growth by moulting or ecdysis.[60] The largest animal phylum belongs here, the Arthropoda, including insects, spiders, crabs, and their kin. All these organisms have a body divided into repeating segments, typically with paired appendages. Two smaller phyla, the Onychophora and Tardigrada, are close relatives of the arthropods and share these traits.

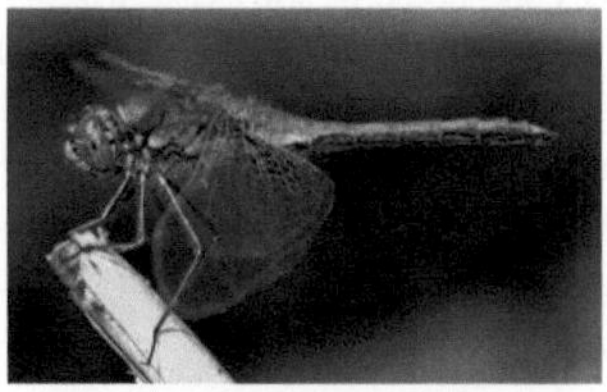

Yellow-winged darter, *Sympetrum flaveolum*

The ecdysozoans also include the Nematoda or roundworms, perhaps the second largest animal phylum. Roundworms are typically microscopic, and occur in nearly every environment where there is water.[61] A number are important parasites.[62] Smaller phyla related to them are the Nematomorpha or horsehair worms, and the Kinorhyncha, Priapulida, and Loricifera. These groups have a reduced coelom, called a pseudocoelom.

The remaining two groups of protostomes are sometimes grouped together as the Spiralia, since in both embryos develop with spiral cleavage.

Platyzoa

The Platyzoa include the phylum Platyhelminthes, the flatworms.[] These were originally considered some of the most primitive Bilateria, but it now appears they developed from more complex ancestors.[63] A number of parasites are included in this group, such as the flukes and tapeworms.[] Flatworms are acoelomates, lacking a body cavity, as are their closest relatives, the microscopic Gastrotricha.[]

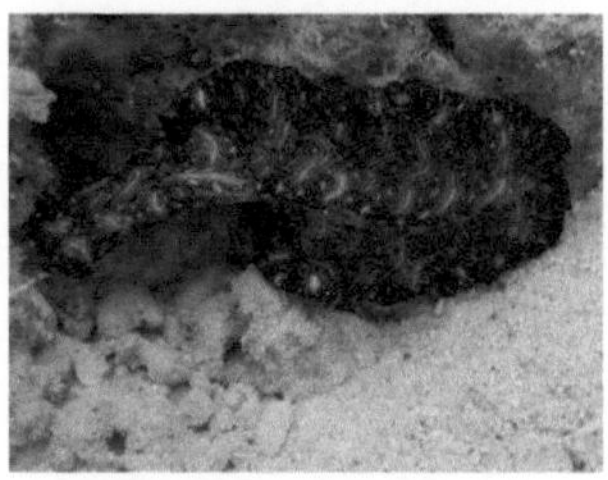

Pseudobiceros bedfordi, (Bedford's flatworm)

The other platyzoan phyla are mostly microscopic and pseudocoelomate. The most prominent are the Rotifera or rotifers, which are common in aqueous environments. They also include the Acanthocephala or spiny-headed worms, the Gnathostomulida, Micrognathozoa, and possibly the Cycliophora.[] These groups share the presence of complex jaws, from which they are called the Gnathifera.

Lophotrochozoa

Roman snail, *Helix pomatia*

The Lophotrochozoa include two of the most successful animal phyla, the Mollusca and Annelida.[64] [65] The former, which is the second-largest animal phylum by number of described species, includes animals such as snails, clams, and squids, and the latter comprises the segmented worms, such as earthworms and leeches. These two groups have long been considered close relatives because of the common presence of trochophore larvae, but the annelids were considered closer to the arthropods because they are both segmented.[66] Now, this is generally considered convergent evolution, owing to many morphological and genetic differences between the two phyla.[67]

The Lophotrochozoa also include the Nemertea or ribbon worms, the Sipuncula, and several phyla that have a ring of ciliated tentacles around the mouth, called a lophophore.[68] These were traditionally grouped together as the lophophorates.[69] but it now appears that the lophophorate group may be paraphyletic,[70] with some closer to the nemerteans and some to the molluscs and annelids.[71] [72] They include the Brachiopoda or lamp shells, which are prominent in the fossil record, the Entoprocta, the Phoronida, and possibly the Bryozoa or moss animals.[73]

Model organisms

Because of the great diversity found in animals, it is more economical for scientists to study a small number of chosen species so that connections can be drawn from their work and conclusions extrapolated about how animals function in general. Because they are easy to keep and breed, the fruit fly *Drosophila melanogaster* and the nematode *Caenorhabditis elegans* have long been the most intensively studied metazoan model organisms, and were among the first life-forms to be genetically sequenced. This was facilitated by the severely reduced state of their genomes, but as many genes, introns, and linkages lost, these ecdysozoans can teach us little about the origins of animals in general. The extent of this type of evolution within the superphylum will be revealed by the crustacean, annelid, and molluscan genome projects currently in progress. Analysis of the starlet sea anemone genome has emphasised the importance of sponges, placozoans, and choanoflagellates, also being sequenced, in explaining the arrival of 1500 ancestral genes unique to the Eumetazoa.[74]

An analysis of the homoscleromorph sponge *Oscarella carmela* also suggests that the last common ancestor of sponges and the eumetazoan animals was more complex than previously assumed.[75]

Other model organisms belonging to the animal kingdom include the house mouse (*Mus musculus*) and zebrafish (*Danio rerio*).

History of classification

Carolus Linnaeus, known as the father of modern taxonomy

Aristotle divided the living world between animals and plants, and this was followed by Carolus Linnaeus (Carl von Linné), in the first hierarchical classification.[] Since then biologists have begun emphasizing evolutionary relationships, and so these groups have been restricted somewhat. For instance, microscopic protozoa were originally considered animals because they move, but are now treated separately.

In Linnaeus's original scheme, the animals were one of three kingdoms, divided into the classes of Vermes, Insecta, Pisces, Amphibia, Aves, and Mammalia. Since then the last four have all been subsumed into a single phylum, the Chordata, whereas the various other forms have been separated out. The above lists represent our current understanding of the group, though there is some variation from source to source.

See also

- Animal colouration
- Ethology
- Fauna
- List of animal names
- List of animals by number of neurons
- List of animals by population
- Lists of animals

References

[23] Begon, M., Townsend, C., Harper, J. (1996). *Ecology: Individuals, populations and communities* (Third edition). Blackwell Science, London. ISBN 0-86542-845-X, ISBN 0-632-03801-2, ISBN 0-632-04393-8.

[24] predation (http://www.britannica.com/search?query=predation). Britannica.com. Retrieved on 2011-11-23.

[32] Monster fish crushed opposition with strongest bite ever (http://www.smh.com.au/news/science/jaws-of-steel-on-this-fish-tank/2006/11/29/1164777657728.html), smh.com.au

[41] The thing about extinction (http://www.bbc.co.uk/lastchancetosee/sites/about/extinction.shtml) retrieved 20 October 2012

[42] Lehmann J, Stadler PF, Krauss V (2012) Near intron pairs and the metazoan tree. *Mol Phylogenet Evol* doi:10.1016/j.ympev.2012.11.012

Bibliography

- Klaus Nielsen. *Animal Evolution: Interrelationships of the Living Phyla* (2nd edition). Oxford University Press, 2001.
- Knut Schmidt-Nielsen. *Animal Physiology: Adaptation and Environment*. (5th edition). Cambridge University Press, 1997.

External links

- *Animal* (http://www.eol.org/pages/1) at the Encyclopedia of Life
- Tree of Life Project (http://tolweb.org/)
- Animal Diversity Web (http://animaldiversity.ummz.umich.edu/site/index.html) – University of Michigan's database of animals, showing taxonomic classification, images, and other information.
- ARKive (http://www.arkive.org) – multimedia database of worldwide endangered/protected species and common species of UK.

- Scientific American Magazine (December 2005 Issue) – Getting a Leg Up on Land (http://www.sciam.com/article.cfm?chanID=sa006&articleID=000DC8B8-EA15-137C-AA1583414B7F0000) About the evolution of four-limbed animals from fish.

Arthropod

Arthropod **Temporal range: Cambrian – Recent**	

Extinct and modern arthropods	
Scientific classification	
Kingdom:	Animalia
Subkingdom:	Eumetazoa
Unranked:	Bilateria
Superphylum:	Ecdysozoa
Phylum:	**Arthropoda** Latreille, 1829
Subphyla and Classes	

- **Subphylum Trilobitomorpha**
  - Trilobita – trilobites (extinct)
- **Subphylum Chelicerata**
  - Arachnida – spiders, scorpions, etc.
  - Merostomata – horseshoe crabs, eurypterids (extinct), etc.
  - Pycnogonida – sea spiders
- **Subphylum Myriapoda**
  - Chilopoda – centipedes
  - Diplopoda – millipedes
  - Pauropoda – sister group to millipedes
  - Symphyla – resemble centipedes
- **Subphylum Crustacea**
  - Branchiopoda – brine shrimp etc.
  - Remipedia – blind crustaceans
  - Cephalocarida – horseshoe shrimp
  - Maxillopoda – barnacles, copepods, fish lice, etc.
  - Ostracoda – seed shrimp
  - Malacostraca – lobsters, crabs, shrimp, etc.
- **Subphylum Hexapoda**
  - Insecta – insects
  - Entognatha – wingless

- Incertae sedis
  - Camptophyllia (extinct)[1]
- **Class Marrellomorpha**
  - Marella (extinct)

An **arthropod** is an invertebrate animal having an exoskeleton (external skeleton), a segmented body, and jointed appendages. Arthropods are members of the phylum **Arthropoda** (from Greek ἄρθρον *árthron*, "joint", and πούς *pous* (gen. *podos*), i.e. "foot" or "leg", which together mean "jointed leg"[]), and include the insects, arachnids, and crustaceans. Arthropods are characterized by their jointed limbs and cuticles, which are mainly made of α-chitin; the cuticles of crustaceans are also biomineralized with calcium carbonate. The rigid cuticle inhibits growth, so arthropods replace it periodically by moulting. The arthropod body plan consists of repeated segments, each with a pair of appendages. Their versatility has enabled them to become the most species-rich members of all ecological guilds in most environments. They have over a million described species, making up more than 80% of all described living animal species, some of which, unlike most animals, are very successful in dry environments. They range in size from microscopic plankton up to forms a few meters long.

Arthropods' primary internal cavity is a hemocoel, which accommodates their internal organs and through which their blood circulates; they have open circulatory systems. Like their exteriors, the internal organs of arthropods are generally built of repeated segments. Their nervous system is "ladder-like", with paired ventral nerve cords running through all segments and forming paired ganglia in each segment. Their heads are formed by fusion of varying numbers of segments, and their brains are formed by fusion of the ganglia of these segments and encircle the esophagus. The respiratory and excretory systems of arthropods vary, depending as much on their environment as on the subphylum to which they belong.

Their vision relies on various combinations of compound eyes and pigment-pit ocelli: in most species the ocelli can only detect the direction from which light is coming, and the compound eyes are the main source of information, but the main eyes of spiders are ocelli that can form images and, in a few cases, can swivel to track prey. Arthropods also have a wide range of chemical and mechanical sensors, mostly based on modifications of the many setae (bristles) that project through their cuticles.

Arthropods' methods of reproduction and development are diverse; all terrestrial species use internal fertilization, but this is often by indirect transfer of the sperm via an appendage or the ground, rather than by direct injection. Aquatic species use either internal or external fertilization. Almost all arthropods lay eggs, but scorpions give birth to live young after the eggs have hatched inside the mother. Arthropod hatchlings vary from miniature adults to grubs and caterpillars that lack jointed limbs and eventually undergo a total metamorphosis to produce the adult form. The level of maternal care for hatchlings varies from nonexistent to the prolonged care provided by scorpions.

The versatility of the arthropod modular body plan has made it difficult for zoologists and paleontologists to classify them and work out their evolutionary ancestry, which dates back to the Cambrian period. From the late 1950s to late 1970s, it was thought that arthropods were polyphyletic, that is, there was no single arthropod ancestor. Now they are generally regarded as monophyletic. Historically, the closest evolutionary relatives of arthropods were considered to be annelid worms, as both groups have segmented bodies. This hypothesis is now largely rejected, with annelids and molluscs forming the superphylum Lophotrochozoa. Many analyses support a placement of arthropods with cycloneuralians (or their constituent clades) in a superphylum Ecdysozoa. Overall however, the basal relationships of Metazoa are not yet well resolved. Likewise, the relationships between various arthropod groups are still actively debated.

Arthropods contribute to the human food supply both directly as food, and more importantly as pollinators of crops. Some specific species are known to spread severe disease to humans, livestock, and crops.

Description

Arthropods are invertebrates with segmented bodies and jointed limbs.[2] The limbs form part of an exoskeleton, which is mainly made of α-chitin, a derivative of glucose.[3] One other group of animals, the tetrapods, has jointed limbs, but tetrapods are vertebrates and therefore have endoskeletons.[] Arthropods bony back evolved so that it became segmented letting it move about more freely and not reducing its protection.

Diversity

One estimate indicates that arthropods have 1,170,000 described species, and account for over 80% of all known living animal species.[4] Another study estimates that there are between 5 to 10 million extant arthropod species, both described and yet to be described.[5] Estimating the total number of living species is extremely difficult because it often depends on a series of assumptions in order to scale up from counts at specific locations to estimates for the whole world. A study in 1992 estimated that there were 500,000 species of animals and plants in Costa Rica alone, of which 365,000 were arthropods. Arthropods are one of the most ubiquitous groups of multicellular organisms found in the biosphere.[]

They are important members of marine, freshwater, land and air ecosystems, and are one of only two major animal groups that have adapted to life in dry environments; the other is amniotes, whose living members are reptiles, birds and mammals.[6] One arthropod sub-group, insects, is the most species-rich member of all ecological guilds in land and fresh-water environments.[] The lightest insects weigh less than 25 micrograms (millionths of a gram),[] while the heaviest weigh over 70 grams (oz).[7] Some living crustaceans are much larger; for example, the legs of the Japanese spider crab may span up to 4 metres (ft).[]

Segmentation

Head

Thorax

Abdomen

Segments and tagmata of an arthropod[6]

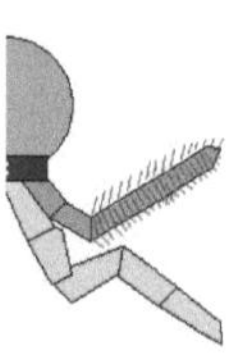

= Body

= Coxa (base)

= Gill branch

// = Gill filaments

= Leg branch

Structure of a biramous appendage[]

The embryos of all arthropods are segmented, built from a series of repeated modules. The last common ancestor of living arthropods probably consisted of a series of undifferentiated segments, each with a pair of appendages that functioned as limbs. However all known living and fossil arthropods have grouped segments into tagmata in which segments and their limbs are specialized in various ways;[6] The three-part appearance of many insect bodies and the two-part appearance of spiders is a result of this grouping;[] in fact there are no external signs of segmentation in mites.[6] Arthropods also have two body elements that are not part of this serially repeated pattern of segments, an acron at the front, ahead of the mouth, and a telson at the rear, behind the anus. The eyes are mounted on the acron.[6]

The original structure of arthropod appendages was probably biramous, with the upper branch acting as a gill while the lower branch was used for walking. In some segments of all known arthropods the appendages have been modified, for example to form gills, mouth-parts, antennae for collecting information,[] or claws for grasping;[8] arthropods are "like Swiss Army knives, each equipped with a unique set of specialized tools."[6] In many arthropods, appendages have vanished from some regions of the body, and it is particularly common for abdominal appendages to have disappeared or be highly modified.[6]

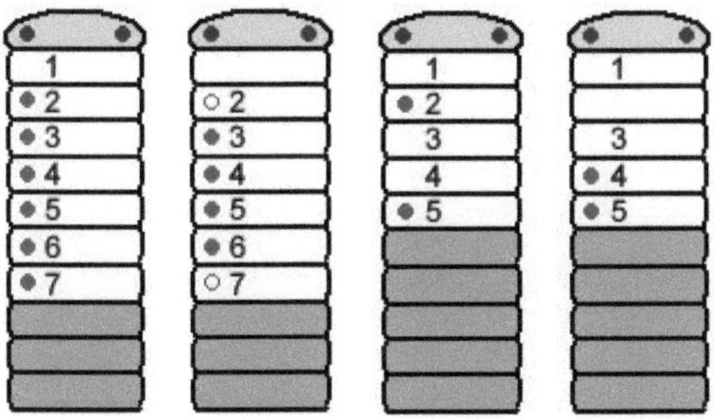

Trilobitomorpha

A

L

L

L

L

L

L

Chelicerata

x

C

P

L

L

L

L

Ci

Crustacea

A

A

Mnd

Mx

Mx

L

L

L

L

L

Tracheata

A

x

Mnd

Mx

Mx

L

L

L

L

= acron

= segments included in head

= body segments

x = lost during development

= eyes

= nephridia

O = nephridia lost during development

A = Antenna

L = Leg

C = Chelicera

P = Pedipalp

Ci = Chilarium

Mnd = Mandible

Mx = Maxilla

The arthropod head problem

The most conspicuous specialization of segments is in the head. The four major groups of arthropods – Chelicerata (includes spiders and scorpions), Crustacea (shrimps, lobsters, crabs, etc.), Tracheata (arthropods that breathe via channels into their bodies; includes insects and myriapods), and the extinct trilobites – have heads formed of various combinations of segments, with appendages that are missing or specialized in different ways.[6] In addition some extinct arthropods, such as *Marrella*, belong to none of these groups, as their heads are formed by their own particular combinations of segments and specialized appendages.[9] Working out the evolutionary stages by which all these different combinations could have appeared is so difficult that it has long been known as "the arthropod head problem".[10] In 1960 R. E. Snodgrass even hoped it would not be solved, as trying to work out solutions was so much fun.[11]

Exoskeleton

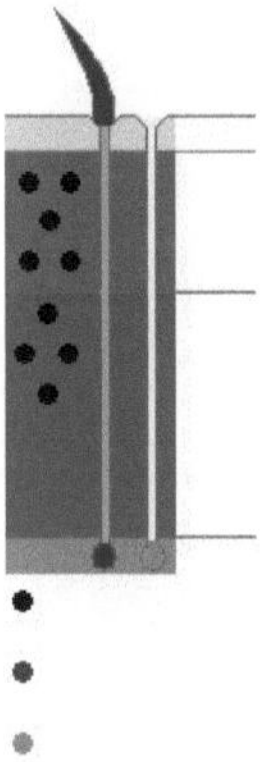

Seta (bristle)

Epicuticle

Exocuticle

Endocuticle

Epidermis

= Biomineralization, only in crustaceans

= Trichogen cell, produces seta

= Gland cell, secretes epicuticle

Structure of arthropod cuticle

Arthropod exoskeletons are made of cuticle, a non-cellular material secreted by the epidermis.[6] Their cuticles vary in the details of their structure, but generally consist of three main layers: the epicuticle, a thin outer waxy coat that moisture-proofs the other layers and gives them some protection; the exocuticle, which consists of chitin and chemically hardened proteins; and the endocuticle, which consists of chitin and unhardened proteins. The exocuticle and endocuticle together are known as the procuticle.[] Each body segment and limb section is encased in hardened cuticle. The joints between body segments and between limb sections are covered by flexible cuticle.[6]

The exoskeletons of most aquatic crustaceans are biomineralized with calcium carbonate extracted from the water. Some terrestrial crustaceans have developed means of storing the mineral, since on land they cannot rely on a steady supply of dissolved calcium carbonate.[12] Biomineralization generally affects the exocuticle and the outer part of the endocuticle.[] Two recent hypotheses about the evolution of biomineralization in arthropods and other groups of animals propose that it provides tougher defensive armor,[] and that it allows animals to grow larger and stronger by providing more rigid skeletons;[] and in either case a mineral-organic composite exoskeleton is cheaper to build than an all-organic one of comparable strength.[] []

The cuticle can have setae (bristles) growing from special cells in the epidermis. Setae are as varied in form and function as appendages. For example, they are often used as sensors to detect air or water currents, or contact with objects; aquatic arthropods use feather-like setae to increase the surface area of swimming appendages and to filter food particles out of water; aquatic insects, which are air-breathers, use thick felt-like coats of setae to trap air, extending the time they can spend under water; heavy, rigid setae serve as defensive spines.[6]

Although all arthropods use muscles attached to the inside of the exoskeleton to flex their limbs, some still use hydraulic pressure to extend them, a system inherited from their pre-arthropod ancestors;[13] for example, all spiders extend their legs hydraulically and can generate pressures up to eight times their resting level.[14]

Moulting

The exoskeleton cannot stretch and thus restricts growth. Arthropods therefore replace their exoskeletons by moulting, or shedding the old exoskeleton after growing a new one that is not yet hardened. Moulting cycles run nearly continuously until an arthropod reaches full size.[15]

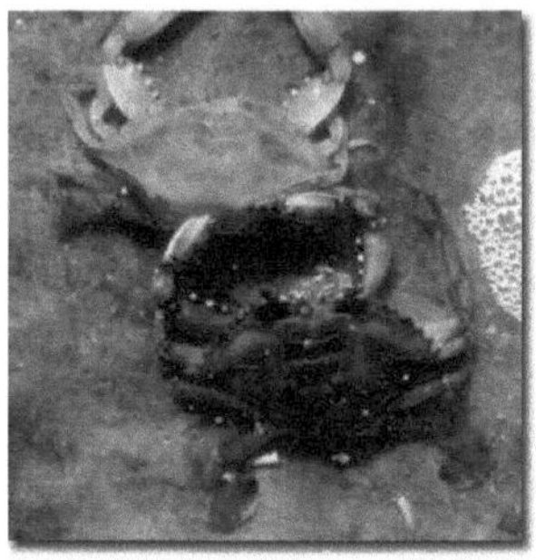

Callinectes sapidus moulting its exoskeleton

In the initial phase of moulting, the animal stops feeding and its epidermis releases moulting fluid, a mixture of enzymes that digests the endocuticle and thus detaches the old cuticle. This phase begins when the epidermis has secreted a new epicuticle to protect it from the enzymes, and the epidermis secretes the new exocuticle while the old cuticle is detaching. When this stage is complete, the animal makes its body swell by taking in a large quantity of water or air, and this makes the old cuticle split along predefined weaknesses where the old exocuticle was thinnest. It commonly takes several minutes for the animal to struggle out of the old cuticle. At this point the new one is wrinkled and so soft that the animal cannot support itself and finds it very difficult to move, and the new endocuticle has not yet formed. The animal continues to pump itself up to stretch the new cuticle as much as possible, then hardens the new exocuticle and eliminates the excess air or water. By the end of this phase the new endocuticle has formed. Many arthropods then eat the discarded cuticle to reclaim its materials.[15]

Because arthropods are unprotected and nearly immobilized until the new cuticle has hardened, they are in danger both of being trapped in the old cuticle and of being attacked by predators. Moulting may be responsible for 80 to 90% of all arthropod deaths.[15]

Internal organs

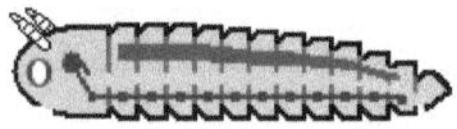

= heart

= gut

= brain, nerve cord, ganglia

O = eye

Basic arthropod body structure

Arthropod bodies are also segmented internally, and the nervous, muscular, circulatory, and excretory systems have repeated components.[6] Arthropods come from a lineage of animals that have a coelom, a membrane-lined cavity between the gut and the body wall that accommodates the internal organs. The strong, segmented limbs of arthropods eliminate the need for one of the coelom's main ancestral functions, as a hydrostatic skeleton, which muscles compress in order to change the animal's shape and thus enable it to move. Hence the coelom of the arthropod is reduced to small areas around the reproductive and excretory systems. Its place is largely taken by a hemocoel, a cavity that runs most of the length of the body and through which blood flows.[16]

Arthropods have open circulatory systems, although most have a few short, open-ended arteries. In chelicerates and crustaceans, the blood carries oxygen to the tissues, while hexapods use a separate system of tracheae. Many crustaceans, but few chelicerates and tracheates, use respiratory pigments to assist oxygen transport. The most common respiratory pigment in arthropods is copper-based hemocyanin; this is used by many crustaceans and a few centipedes. A few crustaceans and insects use iron-based hemoglobin, the respiratory pigment used by vertebrates. As with other invertebrates the respiratory pigments of those arthropods that have them are generally dissolved in the blood and rarely enclosed in corpuscles as they are in vertebrates.[16]

The heart is typically a muscular tube that runs just under the back and for most of the length of the hemocoel. It contracts in ripples that run from rear to front, pushing blood forwards. Sections not being squeezed by the heart muscle are expanded either by elastic ligaments or by small muscles, in either case connecting the heart to the body wall. Along the heart run a series of paired ostia, non-return valves that allow blood to enter the heart but prevent it from leaving before it reaches the front.[16]

Arthropods have a wide variety of respiratory systems. Small species often do not have any, since their high ratio of surface area to volume enables simple diffusion through the body surface to supply enough oxygen. Crustacea usually have gills that are modified appendages. Many arachnids have book lungs. Tracheae, systems of branching tunnels that run from the openings in the body walls, deliver oxygen directly to individual cells in many insects, myriapods and arachnids.[17]

Living arthropods have paired main nerve cords running along their bodies below the gut, and in each segment the cords form a pair of ganglia from which sensory and motor nerves run to other parts of the segment. Although the pairs of ganglia in each segment often appear physically fused, they are connected by commissures (relatively large bundles of nerves), which give arthropod nervous systems a characteristic "ladder-like" appearance. The brain is in the head, encircling and mainly *above* the esophagus. It consists of the fused ganglia of the acron and one or two of the foremost segments that form the head – a total of three pairs of ganglia in most arthropods, but only two in chelicerates, which do not have antennae or the ganglion connected to them. The ganglia of other head segments are often close to the brain and function as part of it. In insects these other head ganglia combine into a pair of subesophageal ganglia, under and behind the esophagus. Spiders take this process a step further, as *all* the segmental ganglia are incorporated into the subesophageal ganglia, which occupy most of the space in the cephalothorax (front "super-segment").[18]

There are two different types of arthropod excretory systems. In aquatic arthropods, the end-product of biochemical reactions that metabolise nitrogen is ammonia, which is so toxic that it needs to be diluted as much as possible with water. The ammonia is then eliminated via any permeable membrane, mainly through the gills. All crustaceans use this system, and its high consumption of water may be responsible for the relative lack of success of crustaceans as land animals.[19] Various groups of terrestrial arthropods have independently developed a different system: the end-product of nitrogen metabolism is uric acid, which can be excreted as dry material; the Malpighian tubule system filters the uric acid and other nitrogenous waste out of the blood in the hemocoel, and dumps these materials into the hindgut, from which they are expelled as feces.[19] Most aquatic arthropods and some terrestrial ones also have organs called nephridia ("little kidneys"), which extract other wastes for excretion as urine.[19]

Senses

The stiff cuticles of arthropods would block out information about the outside world, except that they are penetrated by many sensors or connections from sensors to the nervous system. In fact, arthropods have modified their cuticles into elaborate arrays of sensors. Various touch sensors, mostly setae, respond to different levels of force, from strong contact to very weak air currents. Chemical sensors provide equivalents of taste and smell, often by means of setae. Pressure sensors often take the form of membranes that function as eardrums, but are connected directly to nerves rather than to auditory ossicles. The antennae of most hexapods include sensor packages that monitor humidity, moisture and temperature.[20]

Most arthropods have sophisticated visual systems that include one or more usually both of compound eyes and pigment-cup ocelli ("little eyes"). In most cases ocelli are only capable of detecting the direction from which light is coming, using the shadow cast by the walls of the cup. However the main eyes of spiders are pigment-cup ocelli that are capable of forming images,[20] and those of jumping spiders can rotate to track prey.[21]

Head of a wasp with three ocelli (centre), and compound eyes at the left and right

Compound eyes consist of fifteen to several thousand independent ommatidia, columns that are usually hexagonal in cross section. Each ommatidium is an independent sensor, with its own light-sensitive cells and often with its own lens and cornea.[20] Compound eyes have a wide field of view, and can detect fast movement and, in some cases, the polarization of light.[22] On the other hand the relatively large size of ommatidia makes the images rather coarse, and compound eyes are shorter-sighted than those of birds and mammals – although this is not a severe disadvantage, as objects and events within 20 centimetres (in) are most important to most arthropods.[20] Several arthropods have color vision, and that of some insects has been studied in detail; for example, the ommatidia of bees contain receptors for both green and ultra-violet.[20]

Most arthropods lack balance and acceleration sensors, and rely on their eyes to tell them which way is up. The self-righting behavior of cockroaches is triggered when pressure sensors on the underside of the feet report no pressure. However many malacostracan crustaceans have statocysts, which provide the same sort of information as the balance and motion sensors of the vertebrate inner ear.[20]

The proprioceptors of arthropods, sensors that report the force exerted by muscles and the degree of bending in the body and joints, are well understood. However, little is known about what other internal sensors arthropods may have.[20]

Reproduction and development

A few arthropods, such as barnacles, are hermaphroditic, that is, each can have the organs of both sexes. However, individuals of most species remain of one sex all their lives.[23] A few species of insects and crustaceans can reproduce by parthenogenesis, for example, without mating, especially if conditions favor a "population explosion". However most arthropods rely on sexual reproduction, and parthenogenetic species often revert to sexual reproduction when conditions become less favorable.[] Aquatic arthropods may breed by external fertilization, as for example frogs also do, or by internal fertilization, where the ova remain in the female's body and the sperm must somehow be inserted. All known terrestrial arthropods use internal fertilization, as unprotected sperm and ova would not survive long in these environments. In a few cases the sperm transfer is direct from the male spider's penis to the female's oviduct, but it is more often indirect. Some crustaceans and spiders use modified appendages to transfer the sperm to the female. On the other hand, many male terrestrial arthropods produce spermatophores, waterproof packets of sperm, which the females take into their bodies. A few such species rely on females to find spermatophores that have already been deposited on the ground, but in most cases males only deposit spermatophores when complex courtship rituals look likely to be successful.[23]

Compsobuthus werneri female with young (white)

Most arthropods lay eggs,[23] but scorpions are viviparous: they produce live young after the eggs have hatched inside the mother, and are noted for prolonged maternal care.[24] Newly born arthropods have diverse forms, and insects alone cover the range of extremes. Some hatch as apparently miniature adults (direct development), and in some cases, such as silverfish, the hatchlings do not feed and may be helpless until after their first moult. Many insects hatch as grubs or caterpillars, which do not have segmented limbs or hardened cuticles, and metamorphose into adult forms by entering an inactive phase in which the larval tissues are broken down and re-used to build the adult body.[25] Dragonfly larvae have the typical cuticles and jointed limbs of arthropods but are flightless water-breathers with extendable jaws.[26] Crustaceans commonly hatch as tiny nauplius larvae that have only three segments and pairs of appendages.[23]

The nauplius larva of a prawn

Evolution

Last common ancestor

The last common ancestor of all arthropods is reconstructed as a modular organism with each module covered by its own sclerite (armor plate) and bearing a pair of biramous limbs.[] Whether the ancestral limb was uniramous or biramous is far from a settled debate, though. This Ur-arthropod had a ventral mouth, pre-oral antennae and dorsal eyes at the front of the body. It was a non-discriminatory sediment feeder, processing whatever sediment came its way for food.[]

Fossil record

Marrella, one of the puzzling arthropods from the Burgess Shale

It has been proposed that the Ediacaran animals *Parvancorina* and *Spriggina*, from around 555 Mya, were arthropods.[27] [28] [29] Small arthropods with bivalve-like shells have been found in Early Cambrian fossil beds dating 541 to 539 [30] million years ago in China.[] [31] The earliest Cambrian trilobite fossils are about 530 million years old, but the class was already quite diverse and worldwide, suggesting that they had been around for quite some time.[] Re-examination in the 1970s of the Burgess Shale fossils from about 505 [32] million years ago identified many arthropods, some of which could not be assigned to any of the well-known groups, and thus intensified the debate about the Cambrian explosion.[33] [] [34] A fossil of *Marrella* from the Burgess Shale has provided the earliest clear evidence of moulting.[35]

The earliest fossil crustaceans date from about 513 [36] million years ago in the Cambrian,[37] and fossil shrimp from about 500 [38] million years ago apparently formed a tight-knit procession across the seabed.[39] Crustacean fossils are common from the Ordovician period onwards.[40] They have remained almost entirely aquatic, possibly because they never developed excretory systems that conserve water.[19]

Arthropods provide the earliest identifiable fossils of land animals, from about 419 [41] million years ago in the Late Silurian, and terrestrial tracks from about 450.0 [42] million years ago appear to have been made by arthropods.[43] Arthropods were well pre-adapted to colonize land, because their existing jointed exoskeletons provided protection against desiccation, support against gravity and a means of locomotion that was not dependent on water.[] Around the same time the aquatic, scorpion-like eurypterids became the largest ever arthropods, some as long as 2.5 metres (ft).[44]

The oldest known arachnid is the trigonotarbid *Palaeotarbus jerami*, from about 420 [45] million years ago in the Silurian period.[46] [47] *Attercopus fimbriunguis*, from 386 [48] million years ago in the Devonian period, bears the earliest known silk-producing spigots, but its lack of spinnerets means it was not one of the true spiders,[49] which first appear in the Late Carboniferous over 299 [50] million years ago.[51] The Jurassic and Cretaceous periods provide a large number of fossil spiders, including representatives of many modern families.[] Fossils of aquatic scorpions with gills appear in the Silurian and Devonian periods, and the earliest fossil of an air-breathing scorpion with book lungs dates from the Early Carboniferous period.[52]

The oldest definitive insect fossil is the Devonian *Rhyniognatha hirsti*, dated at 396 to 407 [53] million years ago, but its mandibles are of a type found only in winged insects, which suggests that the earliest insects appeared in the Silurian period.[] The Mazon Creek lagerstätten from the Late Carboniferous, about 300 [54] million years ago, include about 200 species, some gigantic by modern standards, and indicate that insects had occupied their main modern ecological niches as herbivores, detritivores and insectivores. Social termites and ants first appear in the Early Cretaceous, and advanced social bees have been found in Late Cretaceous rocks but did not become abundant until the Middle Cenozoic.[55]

Evolutionary family tree

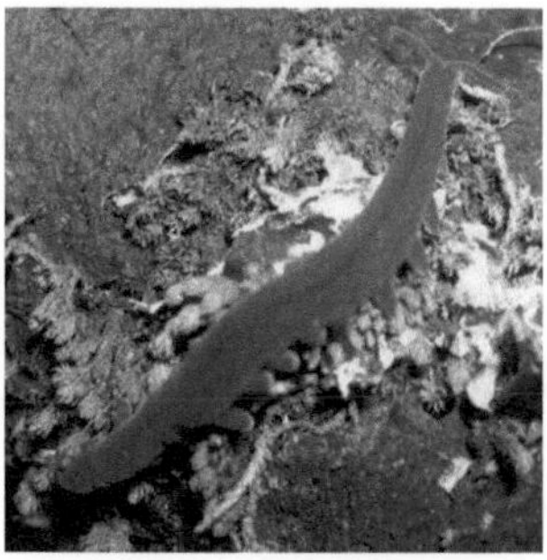

The velvet worm (Onychophora) is closely related to arthropods[]

From 1952 to 1977, Sidnie Manton and others argued that arthropods are polyphyletic, in other words, they do not share a common ancestor that was itself an arthropod. Instead, they proposed that three separate groups of "arthropods" evolved separately from common worm-like ancestors: the chelicerates, including spiders and scorpions; the crustaceans; and the uniramia, consisting of onychophorans, myriapods and hexapods. These arguments usually bypassed trilobites, as the evolutionary relationships of this class were unclear. Proponents of polyphyly argued the following: that the similarities between these groups are the results of convergent evolution, as natural consequences of having rigid, segmented exoskeletons; that the three groups use different chemical means of hardening the cuticle; that there were significant differences in the construction of their compound eyes; that it is hard to see how such different configurations of segments and appendages in the head could have evolved from the same ancestor; and that crustaceans have biramous limbs with separate gill and leg branches, while the other two groups have uniramous limbs in which the single branch serves as a leg.[56]

onychophorans,
including *Aysheaia* and *Peripatus*

armored lobopods,
including *Hallucigenia* and *Microdictyon*

anomalocarid-like taxa,
including modern tardigrades as
well as extinct animals like
Kerygmachela and *Opabinia*

Anomalocaris

arthropods,
including living groups and
extinct forms such as trilobites

Simplified summary of Budd's "broad-scale" cladogram (1996)[]

Further analysis and discoveries in the 1990s reversed this view, and led to acceptance that arthropods are monophyletic, in other words they do share a common ancestor that was itself an arthropod.[57] [58] For example Graham Budd's analyses of *Kerygmachela* in 1993 and of *Opabinia* in 1996 convinced him that these animals were similar to onychophorans and to various Early Cambrian "lobopods", and he presented an "evolutionary family tree" that showed these as "aunts" and "cousins" of all arthropods.[] [] These changes made the scope of the term "arthropod" unclear, and Claus Nielsen proposed that the wider group should be labelled "Panarthropoda" ("all the arthropods") while the animals with jointed limbs and hardened cuticles should be called "Euarthropoda" ("true arthropods").[59]

A contrary view was presented in 2003, when Jan Bergström and Xian-Guang Hou argued that, if arthropods were a "sister-group" to any of the anomalocarids, they must have lost and then re-evolved features that were well-developed in the anomalocarids. The earliest known arthropods ate mud in order to extract food particles from it, and possessed variable numbers of segments with unspecialized appendages that functioned as both gills and legs. Anomalocarids were, by the standards of the time, huge and sophisticated predators with specialized mouths and grasping appendages, fixed numbers of segments some of which were specialized, tail fins, and gills that were very different from those of arthropods. This reasoning implies that *Parapeytoia*, which has legs and a backward-pointing mouth like that of the earliest arthropods, is a more credible closest relative of arthropods than is *Anomalocaris*.[60] In 2006, they suggested that arthropods were more closely related to lobopods and tardigrades than to anomalocarids.[61]

Lophotrochozoa (annelids, molluscs, brachiopods, etc.)

Nematoida (nematodes and close relatives)

Loricifera

Scalidophora (priapulids and Kinorhyncha)

Onychophorans

Tardigrades

Chelicerates

Protostomes

Ecdysozoa

Euthycarcinoids

Panarthropoda

Myriapods

Euarthropoda

Mandibulata

Crustaceans

Hexapods

Relationships of Ecdysozoa to each other and to annelids, etc.,[] including euthycarcinoids[]

Higher up the "family tree", the Annelida have traditionally been considered the closest relatives of the Panarthropoda, since both groups have segmented bodies, and the combination of these groups was labelled Articulata. There had been competing proposals that arthropods were closely related to other groups such as nematodes, priapulids and tardigrades, but these remained minority views because it was difficult to specify in detail the relationships between these groups.

In the 1990s, molecular phylogenetic analyses of DNA sequences produced a coherent scheme showing arthropods as members of a superphylum labelled Ecdysozoa ("animals that moult"), which contained nematodes, priapulids and tardigrades but excluded annelids. This was backed up by studies of the anatomy and development of these animals, which showed that many of the features that supported the Articulata hypothesis showed significant differences between annelids and the earliest Panarthropods in their details, and some were hardly present at all in arthropods. This hypothesis groups annelids with molluscs and brachiopods in another superphylum, Lophotrochozoa.

If the Ecdysozoa hypothesis is correct, then segmentation of arthropods and annelids either has evolved convergently or has been inherited from a much older ancestor and subsequently lost in several other lineages, such as the non-arthropod members of the Ecdysozoa.[62] []

Classification

Arthropods are typically classified into five subphyla, of which one is extinct:[63]

1. **Trilobites** are a group of formerly numerous marine animals that disappeared in the Permian–Triassic extinction event, though they were in decline prior to this killing blow, having been reduced to one order in the Late Devonian extinction.
2. **Chelicerates** include spiders, mites, scorpions and related organisms. They are characterised by the presence of chelicerae, appendages just above / in front of the mouth. Chelicerae appear in scorpions as tiny claws that they use in feeding, but those of spiders have developed as fangs that inject venom.
3. **Myriapods** comprise millipedes, centipedes, and their relatives and have many body segments, each bearing one or two pairs of legs. They are sometimes grouped with the hexapods.
4. **Crustaceans** are primarily aquatic (a notable exception being woodlice) and are characterised by having biramous appendages. They include lobsters, crabs, barnacles, crayfish, shrimp and many others.
5. **Hexapods** comprise insects and three small orders of insect-like animals with six thoracic legs. They are sometimes grouped with the myriapods, in a group called Uniramia, though genetic evidence tends to support a closer relationship between hexapods and crustaceans.

Aside from these major groups, there are also a number of fossil forms, mostly from the Early Cambrian, which are difficult to place, either from lack of obvious affinity to any of the main groups or from clear affinity to several of them. *Marrella* was the first one to be recognized as significantly different from the well-known groups.[9]

The phylogeny of the major extant arthropod groups has been an area of considerable interest and dispute.[64] Recent studies strongly suggest that Crustacea, as traditionally defined, is paraphyletic, with Hexapoda having evolved from within it,[] [] so that Crustacea and Hexapoda form a clade, Pancrustacea. The position of Myriapoda, Chelicerata and Pancrustacea remains unclear as of April 2012. In some studies, Myriapoda is grouped with Chelicerata (forming Myriochelata);[] [65] in other studies, Myriapoda is grouped with Pancrustacea (forming Mandibulata),[] or Myriapoda may be sister to Chelicerata plus Pancrustacea.[]

Onchyophora

Tardigrada

Chelicerata

Myriapoda

Ostracoda, Branchiura, Pentastomida, Mystacocarida

Panarthropoda

Euarthropoda

Branchiopoda, Copepoda, Malacostraca, Thecostraca

Mandibulata

Pancrustacea

Remipedia, Cephalocarida

Hexapoda

Phylogenetic relationships of the major extant arthropod groups according to Regier et al. (2010);[] traditional subphyla in bold

The placement of the extinct trilobites is also a frequent subject of dispute.[] One of the newer hypotheses is that the chelicerae have originated from the same pair of appendages that evolved into antennae in the ancestors of Mandibulata, which would place trilobites, which had antennae, closer to Mandibulata than Chelicerata.[66]

Since the International Code of Zoological Nomenclature recognises no priority above the rank of family, many of the higher-level groups can be referred to by a variety of different names.[]

Interaction with humans

Crustaceans such as crabs, lobsters, crayfish, shrimp and prawns have long been part of human cuisine, and are now farmed on a large commercial scale.[67] Insects and their grubs are at least as nutritious as meat, and are eaten both raw and cooked in many cultures, excluding most European, Hindu and Islamic cultures.[68] [69] Cooked tarantulas are considered a delicacy in Cambodia,[70] [71] [72] and by the Piaroa Indians of southern Venezuela, after the highly irritant hairs – the spider's main defense system – are removed.[73] Humans also unintentionally eat arthropods in other foods,[74] and food safety regulations lay down acceptable contamination levels for different kinds of food material.[75] [76] The intentional cultivation of arthropods and other small animals for human food, referred to as minilivestock, is now emerging in animal husbandry as an ecologically sound concept.[77]

Insects and scorpions on sale in a food stall in Bangkok

However, the greatest contribution of arthropods to human food supply is by pollination: a 2008 study examined the 100 crops that FAO lists as grown for food, and estimated pollination's economic value as €153 billion, or 9.5% of the value of world agricultural production used for human food in 2005.[78] Besides pollinating, bees produce honey, which is the basis of a rapidly growing industry and international trade.[79]

The red dye cochineal, produced from a Central American species of insect, was economically important to the Aztecs and Mayans,[80] and while the region was under Spanish control, becoming Mexico's second most-lucrative export;[81] and it is now regaining some of the ground it lost to synthetic competitors.[82] The blood of horseshoe crabs contains a clotting agent Limulus Amebocyte Lysate which is now used to test that antibiotics and kidney machines are free of dangerous bacteria, and to detect spinal meningitis and some cancers.[83] Forensic entomology uses evidence provided by arthropods to establish the time and sometimes the place of death of a human, and in some cases the cause.[84] Recently insects have also gained attention as potential sources of drugs and other medicinal substances.[85]

The relative simplicity of the arthropods' body plan, allowing them to move on a variety of surfaces both on land and in water, have made them useful as models for robotics. The redundancy provided by segments allows arthropods and biomimetic robots to move normally even with damaged or lost appendages.[86] [87]

Diseases transmitted by insects

Disease[]	Insect	Cases per year	Deaths per year
Malaria	*Anopheles* mosquito	267 M	1 to 2 M
Yellow fever	*Aedes* mosquito	4,432	1,177
Filariasis	*Culex* mosquito	250 M	unknown

Although arthropods are the most numerous phylum on Earth, and thousands of arthropod species are venomous, they inflict relatively few serious bites and stings on humans. Far more serious are the effects on humans of diseases carried by blood-sucking insects. Other blood-sucking insects infect livestock with diseases that kill many animals and greatly reduce the usefulness of others.[] Ticks can cause tick paralysis and several parasite-borne diseases in humans.[88] A few of the closely related mites also infest humans, causing intense itching,[89] and others cause allergic diseases, including hay fever, asthma and eczema.[90]

Many species of arthropods, principally insects but also mites, are agricultural and forest pests.[] [91] The mite *Varroa destructor* has become the largest single problem faced by beekeepers worldwide.[92] Efforts to control arthropod pests by large-scale use of pesticides have caused long term effects on human health and on biodiversity.[93] Increasing arthropod resistance to pesticides has led to the development of integrated pest management using a wide range of measures including biological control.[] Predatory mites may be useful in controlling some mite pests.[] [94]

Notes

[6] Ruppert, Fox & Barnes (2004), pp. 518–522

[9] Summarised in Gould (1990), pp. 107–121.

[11] "It would be too bad if the question of head segmentation ever should be finally settled; it has been for so long such fertile ground for theorizing that arthropodists would miss it as a field for mental exercise."

[15] Ruppert, Fox & Barnes (2004), pp. 523–524

[16] Ruppert, Fox & Barnes (2004), pp. 527–528

[17] Ruppert, Fox & Barnes (2004), pp. 530, 733

[18] Ruppert, Fox & Barnes (2004), pp. 531–532

[19] Ruppert, Fox & Barnes (2004), pp. 529–530

[20] Ruppert, Fox & Barnes (2004), pp. 532–537

[21] Ruppert, Fox & Barnes (2004), pp. 578–580

[23] Ruppert, Fox & Barnes (2004), pp. 537–539

[30] http://toolserver.org/~verisimilus/Timeline/Timeline.php?Ma=541–539

[32] http://toolserver.org/~verisimilus/Timeline/Timeline.php?Ma=505

[33] Whittington, H. B. (1979). Early arthropods, their appendages and relationships. In M. R. House (Ed.), The origin of major invertebrate groups (pp. 253–268). The Systematics Association Special Volume, 12. London: Academic Press.

[34] Gould (1990)

[36] http://toolserver.org/~verisimilus/Timeline/Timeline.php?Ma=513

[38] http://toolserver.org/~verisimilus/Timeline/Timeline.php?Ma=500

[41] http://toolserver.org/~verisimilus/Timeline/Timeline.php?Ma=419

[42] http://toolserver.org/~verisimilus/Timeline/Timeline.php?Ma=450

[45] http://toolserver.org/~verisimilus/Timeline/Timeline.php?Ma=420

[47] The fossil was originally named *Eotarbus* but was renamed when it was realized that a Carboniferous arachnid had already been named *Eotarbus*.

[48] http://toolserver.org/~verisimilus/Timeline/Timeline.php?Ma=386

[50] http://toolserver.org/~verisimilus/Timeline/Timeline.php?Ma=299

[53] http://toolserver.org/~verisimilus/Timeline/Timeline.php?Ma=396–407

[54] http://toolserver.org/~verisimilus/Timeline/Timeline.php?Ma=300

[57] The book is

[66] Fossil Focus: Chelicerata (http://www.palaeontologyonline.com/articles/2011/fossil-focus-chelicerata/)

[75] For a mention of insect contamination in an international food quality standard, see sections 3.1.2 and 3.1.3 of Codex 152 of 1985 of the Codex Alimentarius.

[76] For examples of quantified acceptable insect contamination levels in food see the last entry (on "Wheat Flour") and the definition of "Extraneous material" in *Codex Alimentarius*, and the standards published by the FDA.

[78] Free summary at

References

External links

- *Arthropod* (http://www.eol.org/pages/164) at the Encyclopedia of Life
- Venomous Arthropods (http://entomology.ifas.ufl.edu/fasulo/vector/chapter_07.htm) chapter in United States Environmental Protection Agency and University of Florida/Institute of Food and Agricultural Sciences National Public Health Pesticide Applicator Training Manual
- Arthropods - Arthropoda (http://www.insectlifeforms.com/Phylums488/JOINTED_LEGGED_ANIMALS_ARTHROPODA_6050_488.aspx) Insect Life Forms

Crustacean

Crustacea **Temporal range: Cambrian to Recent**	

Abludomelita obtusata, an amphipod	
Scientific classification	
Kingdom:	Animalia
Phylum:	Arthropoda
Subphylum:	**Crustacea** Brünnich, 1772
Classes & Subclasses	

Thylacocephala?
Branchiopoda
    Phyllopoda
    Sarsostraca
Remipedia
Cephalocarida
Maxillopoda
    Thecostraca
    Tantulocarida
    Branchiura
    Pentastomida
    Mystacocarida
    Copepoda
Ostracoda
    Myodocopa
    Podocopa
Malacostraca
    Phyllocarida
    Hoplocarida
    Eumalacostraca	

Crustaceans (**Crustacea**) form a very large group of arthropods, usually treated as a subphylum, which includes such familiar animals as crabs, lobsters, crayfish, shrimp, krill and barnacles. The 67,000 described species range in size from *Stygotantulus stocki* at 0.1 mm (in), to the Japanese spider crab with a leg span of up to 12.5 ft (m) and a mass of 44 lb (kg). Like other arthropods, crustaceans have an exoskeleton, which they moult to grow. They are distinguished from other groups of arthropods, such as insects, myriapods and chelicerates, by the possession of biramous (two-parted) limbs, and by the nauplius form of the larvae.

Most crustaceans are free-living aquatic animals, but some are terrestrial (e.g. woodlice), some are parasitic (e.g. Rhizocephala, fish lice, tongue worms) and some are sessile (e.g. barnacles). The group has an extensive fossil record, reaching back to the Cambrian, and includes living fossils such as *Triops cancriformis*, which has existed apparently unchanged since the Triassic period. More than 10 million tons of crustaceans are produced by fishery or farming for human consumption, the majority of it being shrimp and prawns. Krill and copepods are not as widely fished, but may be the animals with the greatest biomass on the planet, and form a vital part of the food chain. The scientific study of crustaceans is known as **carcinology** (alternatively, *malacostracology*, *crustaceology* or *crustalogy*), and a scientist who works in carcinology is a carcinologist.

Structure

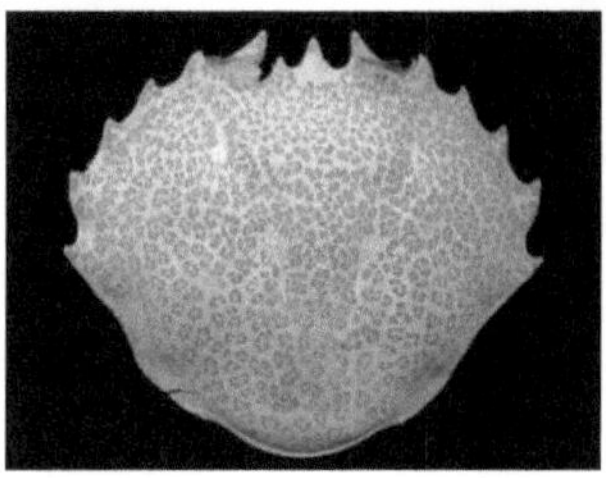

A shed carapace of a lady crab, part of the hard exoskeleton

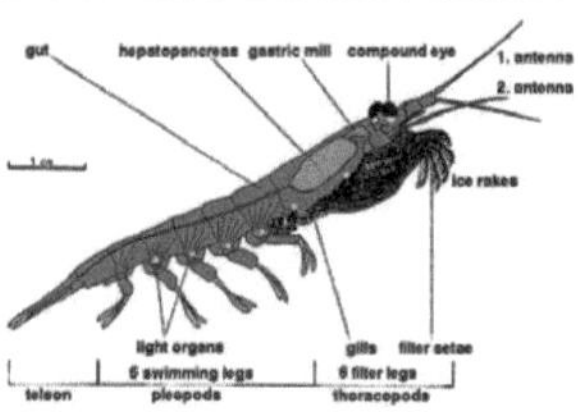

Body structure of a typical crustacean – krill

The body of a crustacean is composed of body segments, which are grouped into three regions: the *cephalon* or head,[] the thorax,[] and the *pleon* or abdomen.[] The head and thorax may be fused together to form a cephalothorax,[1] which may be covered by a single large carapace.[2] The crustacean body is protected by the hard exoskeleton, which must be moulted for the animal to grow. The shell around each somite can be divided into a dorsal tergum, ventral sternum and a lateral pleuron. Various parts of the exoskeleton may be fused together.[] :289

Each somite, or body segment can bear a pair of appendages: on the segments of the head, these include two pairs of antennae, the mandibles and maxillae;[] the thoracic segments bear legs, which may be specialised as pereiopods (walking legs) and maxillipeds (feeding legs).[] The abdomen bears pleopods,[] and ends in a telson, which bears the anus, and is often flanked by uropods to form a tail fan.[] The number and variety of appendages in different crustaceans may be partly responsible for the group's success.[3] Crustacean appendages are typically biramous, meaning they are divided into two parts; this includes the second pair of antennae, but not the first, which is uniramous.[4] It is unclear whether the biramous condition is a derived state which evolved in crustaceans, or whether the second branch of the limb has been lost in all other groups. Trilobites, for instance, also possessed biramous appendages.[5]

The main body cavity is an open circulatory system, where blood is pumped into the haemocoel by a heart located near the dorsum.[6] Malacostraca have haemocyanin as the oxygen-carrying pigment, while copepods, ostracods, barnacles and branchiopods have haemoglobins.[7] The alimentary canal consists of a straight tube that often has a gizzard-like "gastric mill" for grinding food and a pair of digestive glands that absorb food; this structure goes in a spiral format.[8] Structures that function as kidneys are located near the antennae. A brain exists in the form of ganglia close to the antennae, and a collection of major ganglia is found below the gut.[]

In many decapods, the first (and sometimes the second) pair of pleopods are specialised in the male for sperm transfer. Many terrestrial crustaceans (such as the Christmas Island red crab) mate seasonally and return to the sea to release the eggs. Others, such as woodlice, lay their eggs on land, albeit in damp conditions. In most decapods, the females retain the eggs until they hatch into free-swimming larvae.[9]

Ecology

The majority of crustaceans are aquatic, living in either marine or freshwater environments, but a few groups have adapted to life on land, such as terrestrial crabs, terrestrial hermit crabs, and woodlice. Marine crustaceans are as ubiquitous in the oceans as insects are on land.[10] [11] The majority of crustaceans are also motile, moving about independently, although a few taxonomic units are parasitic and live attached to their hosts (including sea lice, fish lice, whale lice, tongue worms, and *Cymothoa exigua*, all of which may be referred to as "crustacean lice"), and adult barnacles live a sessile life – they are attached headfirst to the substrate and cannot move independently. Some branchiurans are able to withstand rapid changes of salinity and will also switch hosts from marine to non-marine species.[] :672 Krill are the bottom layer and the most important part of the food chain in Antarctic animal communities.[12] :64 Some crustaceans are significant invasive species, such as the Chinese mitten crab, *Eriocheir sinensis*,[13] and the Asian shore crab, *Hemigrapsus sanguineus*.[14]

Life cycle

Mating system

Eggs of *Potamon fluviatile*, a freshwater crab

The majority of crustaceans have separate sexes, and reproduce sexually.[] A small number are hermaphrodites, including barnacles, remipedes,[15] and Cephalocarida.[] Some may even change sex during the course of their life.[] Parthenogenesis is also widespread among crustaceans, where viable eggs are produced by a female without needing fertilisation by a male.[] This occurs in many branchiopods, some ostracods, some isopods, and certain "higher" crustaceans, such as the *Marmorkrebs* crayfish.

Eggs

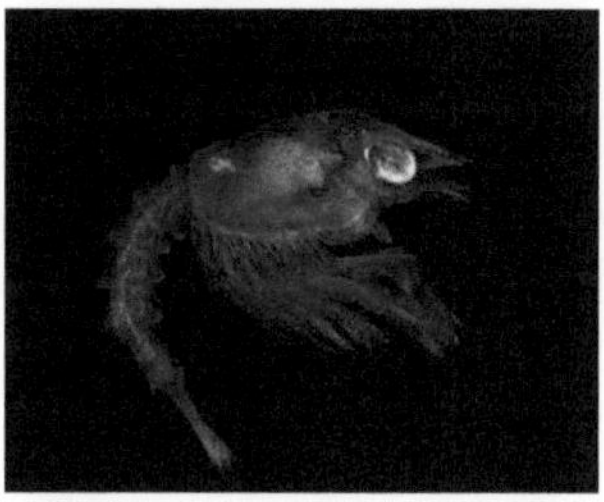

Zoea larva of the European lobster, *Homarus gammarus*

In many groups of crustaceans, the fertilised eggs are simply released into the water column, while others have developed a number of mechanisms for holding on to the eggs until they are ready to hatch. Most decapods carry the eggs attached to the pleopods, while peracarids, notostracans, anostracans, and many isopods form a brood pouch from the carapace and thoracic limbs.[] Female Branchiura do not carry eggs in external ovisacs but attach them in rows to rocks and other objects.[] :788 Most leptostracans and krill carry the eggs between their thoracic limbs; some copepods carry their eggs in special thin-walled sacs, while others have them attached together in long, tangled strings.[]

Larvae

Crustaceans exhibit a number of larval forms, of which the earliest and most characteristic is the nauplius. This has three pairs of appendages, all emerging from the young animal's head, and a single naupliar eye. In most groups, there are further larval stages, including the zoea (pl. zoeæ or zoeas[16]). This name was given to it when naturalists believed it to be a separate species.[17] It follows the nauplius stage and precedes the post-larva. Zoea larvae swim with their thoracic appendages, as opposed to nauplii, which use cephalic appendages, and megalopa, which use abdominal appendages for swimming. It often has spikes on its carapace, which may assist these small organisms in maintaining directional swimming.[18] In many decapods, due to their accelerated development, the zoea is the first larval stage. In some cases, the zoea stage is followed by the mysis stage, and in others, by the megalopa stage, depending on the crustacean group involved.

Classification

The name "crustacean" dates from the earliest works to describe the animals, including those of Pierre Belon and Guillaume Rondelet, but the name was not used by some later authors, including Carl Linnaeus, who included crustaceans among the "Aptera" in his *Systema Naturae*.[] The earliest nomenclaturally valid work to use the name "Crustacea" was Morten Thrane Brünnich's *Zoologiæ Fundamenta* in 1772,[19] although he also included chelicerates in the group.[]

The subphylum Crustacea comprises almost 67,000 described species,[20] which is thought to be just $^{1}\!/_{10}$ to $^{1}\!/_{100}$ of the total number as the majority of species remain as yet undiscovered.[21] Although most crustaceans are small, their morphology varies greatly and includes both the largest arthropod in the world – the Japanese spider crab with a leg span of metres (12 ft)[22] – and the smallest, the 1-micrometre-long (0.00004 in) *Stygotantulus stocki*.[23] Despite their diversity of form, crustaceans are united by the special larval form known as the nauplius.

The exact relationships of the Crustacea to other taxa are not completely settled as of April 2012. Studies based on morphology led to the Pancrustacea hypothesis,[24] in which Crustacea and Hexapoda (insects and allies) are sister groups. More recent studies using DNA sequences suggest that Crustacea is paraphyletic, with the hexapods nested within a larger Pancrustacea clade.[] []

Although the classification of crustaceans has been quite variable, the system used by Martin and Davis[] largely supersedes earlier works. Mystacocarida and Branchiura, here treated as part of Maxillopoda, are sometimes treated as their own classes. Six classes are usually recognised:

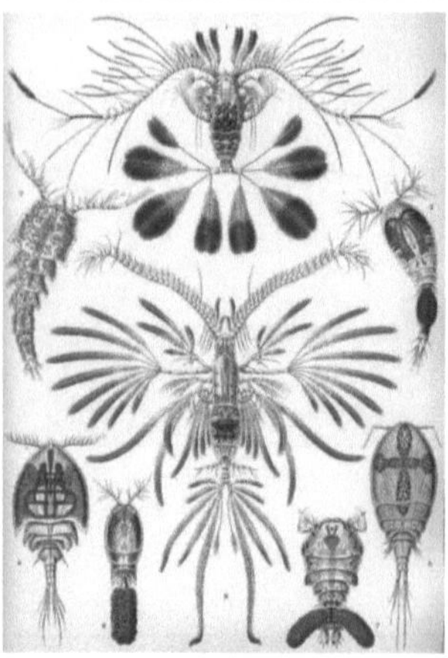

Copepods, from Ernst Haeckel's 1904 work *Kunstformen der Natur*

Decapods, from Ernst Haeckel's 1904 work *Kunstformen der Natur*

Class	Members	Orders	Photo

Branchiopoda	brine shrimp Cladocera *Triops*	Anostraca Lipostraca Notostraca Laevicaudata Spinicaudata Cyclestherida Cladocera	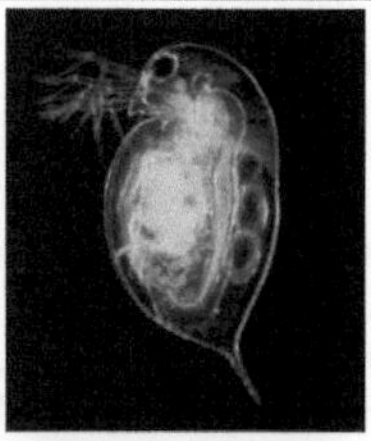 *Daphnia pulex* (Cladocera)
Remipedia		Nectiopoda	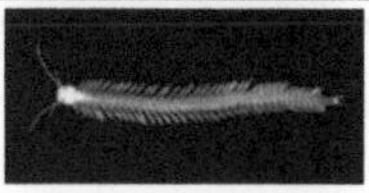 *Speleonectes tanumekes* (Speleonectidae)
Cephalocarida	horseshoe shrimp	Brachypoda	
Maxillopoda	barnacles copepods	Calanoida Pedunculata Sessilia *c.* 20 others	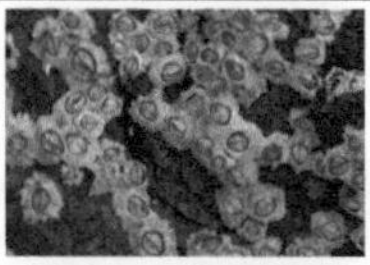 *Chthamalus stellatus* (Sessilia)
Ostracoda	ostracods	Myodocopida Halocyprida Platycopida Podocopida	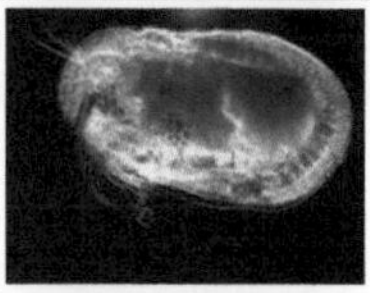 Cylindroleberididae
Malacostraca	crabs lobsters shrimp krill mantis shrimp woodlice sandhoppers *etc.*	Decapoda Isopoda Amphipoda Stomatopoda *c.* 12 others	 *Gammarus roeseli* (Amphipoda)

Fossil record

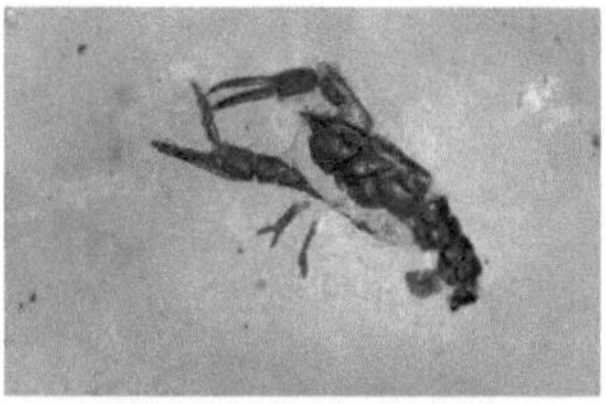

Eryma mandelslohi, a fossil decapod from the Jurassic of Bissingen an der Teck, Germany

Crustaceans have a rich and extensive fossil record, which begins with animals such as *Canadaspis* and *Perspicaris* from the Middle Cambrian age Burgess Shale.[] [] Most of the major groups of crustaceans appear in the fossil record before the end of the Cambrian, namely the Branchiopoda, Maxillopoda (including barnacles and tongue worms) and Malacostraca; there is some debate as to whether or not Cambrian animals assigned to Ostracoda are truly ostracods, which would otherwise start in the Ordovician.[25] The only classes to appear later are the Cephalocarida,[26] which have no fossil record, and the Remipedia, which were first described from the fossil *Tesnusocaris goldichi*, but do not appear until the Carboniferous.[27] Most of the early crustaceans are rare, but fossil crustaceans become abundant from the Carboniferous onwards.[]

Within the Malacostraca, no fossils are known for krill,[28] while both Hoplocarida and Phyllopoda contain important groups that are now extinct as well as extant members (Hoplocarida: mantis shrimp are extant, while Aeschronectida are extinct;[29] Phyllopoda: Canadaspidida are extinct, while Leptostraca are extant[]). Cumacea and Isopoda are both known from the Carboniferous,[30] [31] as are the first true mantis shrimp.[32] In the Decapoda, prawns and polychelids appear in the Triassic,[33] [34] and shrimp and crabs appear in the Jurassic;[35] [] however, the great radiation of crustaceans occurred in the Cretaceous, particularly in crabs, and may have been driven by the adaptive radiation of their main predators, bony fish.[] The first true lobsters also appear in the Cretaceous.[36]

Norway lobsters on sale at a Spanish market

Consumption by humans

Many crustaceans are consumed by humans, and nearly 10,700,000 tons were produced in 2007; the vast majority of this output is of decapod crustaceans: crabs, lobsters, shrimp and prawns.[] Over 60% by weight of all crustaceans caught for consumption are shrimp and prawns, and nearly 80% is produced in Asia, with China alone producing nearly half the world's total.[] Non-decapod crustaceans are not widely consumed, with only 118,000 tons of krill being caught,[] despite krill having one of the greatest biomasses on the planet.[37]

References

External links

- Crustacea.net, an online resource on the biology of crustaceans (http://www.crustacea.net)
- Crustacea (http://crustacea.nhm.org/): Natural History Museum of Los Angeles County
- Crustacea (http://www.tolweb.org/Crustacea/2529): Tree of Life Web Project
- The Crustacean Society (http://web.vims.edu/tcs/?svr=www)
- Natural History Collections: Crustacea (http://www.nhc.ed.ac.uk/index.php?page=24.25.312.330.363): University of Edinburgh
- Crustaceans (Crustacea) on the shore of Singapore (http://www.wildsingapore.com/wildfacts/crustacea/crustacea.htm)
- Crustacea(crabs, lobsters, shrimps, prawns, barnacles) (http://www.biodiversityexplorer.org/mm/crustacea/): Biodiversity Explorer
-

Malacostraca

Malacostraca	
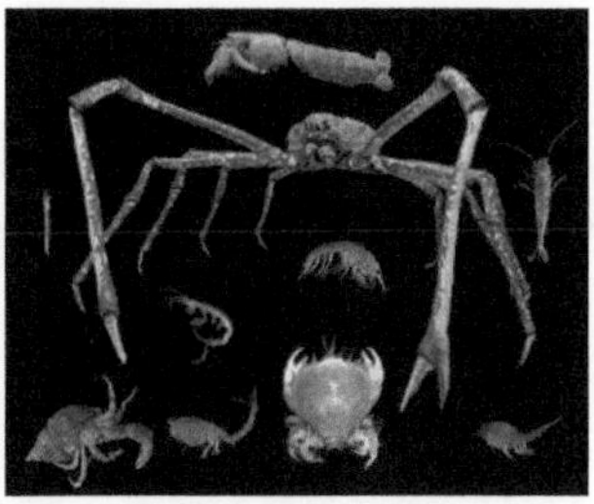	
Scientific classification	
Kingdom:	Animalia
Phylum:	Arthropoda
Subphylum:	Crustacea
Class:	**Malacostraca** Latreille, 1802
Subclasses	
• Eumalacostraca • Hoplocarida • Phyllocarida See text for orders.	

Malacostraca is the largest of the six classes of crustaceans, containing over 25,000 extant species, divided among 16 orders. Its members display a greater diversity of body forms than any other class of animals, and include crabs, lobsters, shrimp, krill, woodlice, scuds (Amphipoda), mantis shrimp and many other less familiar animals. They are abundant in all marine environments and have also colonised freshwater and terrestrial habitats. They are united by a common *Bauplan*, comprising 20 body segments (rarely 21), divided into a head, thorax and abdomen.

Ecology

Grapsus grapsus, a terrestrial crab

Malacostracans live in a wide range of marine and freshwater habitats, and three orders have terrestrial members: Amphipoda (Talitridae), Isopoda (Oniscidea, the woodlice) and Decapoda (terrestrial hermit crabs, crabs of the families Ocypodidae, Gecarcinidae and Grapsidae, and terrestrial crayfish).[1] They are abundant in all marine ecosystems, and most species are scavengers, although some, such as the porcelain crabs, are filter feeders, and some, such as mantis shrimp, are carnivores.[]

Etymology

The name Malacostraca comes from the Greek roots μαλακός (*malakós*, meaning "soft") and ὄστρακον (*óstrakon*, meaning "shell").[2] The name is misleading, since the shell is only soft immediately after moulting, and is usually hard.[] Malacostracans are sometimes contrasted with entomostracans, a name applied to all crustaceans outside the Malacostraca, and named after the obsolete taxon Entomostraca.[]

Description

The Class Malacostraca includes over 25,000 species,[] and "arguably ... contains a greater diversity of body forms than any other class in the animal kingdom".[] Its members are characterised by the presence of three tagmata – a five-segmented head, an eight-segmented thorax and an abdomen with six segments, except in the Leptostraca, which retain the ancestral condition of seven abdominal segments.[] This arrangement is known as the "caridoid facies", a term coined by William Thomas Calman in 1909. Each body segment bears a pair of jointed appendages, although these may be lost secondarily.[]

Leptostraca such as *Nebalia bipes* retain the primitive condition of having seven abdominal segments.

Tagmata

The head bears two pairs of antennae, the first of which is biramous and the second uniramous, and two pairs of maxillae. There is usually a pair of stalked compound eyes, although these may be sessile, reduced or lost.[]

Up to eight thoracic segments may be fused with the head to form a cephalothorax, and up to three pairs of appendages may be modified as maxillipeds (accessory mouthparts).[] A carapace may be absent, present, or secondarily lost, and may cover from two thoracic segments to the entire thorax and some of the abdomen.[]

Each segment of the abdomen except the last carries a pair of pleopods. The appendages of the last segment are typically flattened into uropods, which together with the terminal telson, make up the "tail fan".[] In Leptostraca, these appendages instead form caudal rami.[3]

Life cycle

Most species of malacostraca have distinct sexes (a phenomenon known as gonochorism) although there are a few species that exhibit hermaphroditism.[] The female genital openings are on the sixth thoracic segment, while the male genital opening is usually on the sixth thoracic segment, but is occasionally on the seventh.[] Each of the thoracic appendages is biramous and also carries a gill.[] The larval stages are often reduced, but where they occur, there is usually a metamorphosis between the larval and the adult form.[]

Classification

Martin and Davis present the following classification of living malacostracans into orders,[] to which extinct orders have been added, indicated by an obelisk (†).

Class **Malacostraca** Latreille, 1802

- Subclass Phyllocarida Packard, 1879
  - † Archaeostraca Claus 1888
  - † Hoplostraca Schram, 1973
  - † Canadaspidida Novožilov *in* Orlov, 1960
  - † Hymenostraca Rolfe, 1969
  - Leptostraca Claus, 1880
- Subclass Hoplocarida Calman, 1904
  - † Aeschronectida Schram, 1969
  - † Archaeostomatopoda Schram, 1969
  - Stomatopoda Latreille, 1817
- Subclass Eumalacostraca Grobben, 1892
  - Superorder Syncarida Packard, 1885
    - † Palaeocaridacea Brooks, 1979
    - Bathynellacea Chappuis, 1915
    - Anaspidacea Calman, 1904
  - Superorder Peracarida Calman, 1904
    - Spelaeogriphacea Gordon, 1957
    - Thermosbaenacea Monod, 1927
    - Lophogastrida Sars, 1870
    - Mysida Haworth, 1825
    - Mictacea Bowman *et al.*, 1985
    - Amphipoda Latreille, 1816
    - Isopoda Latreille, 1817
    - Tanaidacea Dana, 1849
    - Cumacea Krøyer, 1846
  - Superorder Eucarida Calman, 1904
    - Euphausiacea Dana, 1852
    - Amphionidacea Williamson, 1973
    - Decapoda Latreille, 1802

Odontodactylus scyllarus (Hoplocarida: Stomatopoda)

Porcellio scaber and *Oniscus asellus* (Peracarida: Isopoda)

Cancer pagurus (Eucarida: Decapoda)

Phylogenetics

While the monophyly of Malacostraca as a whole is widely supported, a number of problems make it difficult to determine the relationships between the orders of Malacostraca. These include differences in rates of evolution in different lineages, different patterns of evolution being apparent in different sources of data, including convergent evolution, and long branch attraction.[4]

Fossil record

The first malacostracans appeared in the Early Cambrian,[5] when animals belonging to the Phyllocarida appeared.[]

References

External links

External identifiers for *Malacostrata*	
Encyclopedia of Life	1157 (http://www.eol.org/pages/1157)
ITIS	89787 (http://www.itis.gov/servlet/SingleRpt/SingleRpt?search_topic=TSN&search_value=89787)
NCBI	6681 (http://www.ncbi.nlm.nih.gov/Taxonomy/Browser/wwwtax.cgi?mode=Info&id=6681)
WoRMS	1071 (http://www.marinespecies.org/aphia.php?p=taxdetails&id=1071)
Also found in: Wikispecies, ADW (http://animaldiversity.ummz.umich.edu/site/accounts/information/Malacostraca.html) ToLWeb (http://tolweb.org/Malacostraca/6253)	

- Malacostraca (http://tolweb.org/Malacostraca/6253), Tree of Life Web Project
- Introduction to the Malacostraca (http://www.ucmp.berkeley.edu/arthropoda/crustacea/malacostraca.html), University of California, Berkeley
- "Malacostraca" (http://www.1911encyclopedia.org/Malacostraca) by Thomas Roscoe Rede Stebbing, Encyclopædia Britannica Eleventh Edition
- Malacostraca (http://paleodb.org/cgi-bin/bridge.pl?action=basicTaxonInfo&taxon_no=22215), The Paleobiology Database
- Malacostraca image key - Guide to the marine zooplankton of south eastern Australia (http://www.tafi.org.au/zooplankton/imagekey/malacostraca/), Tasmanian Aquaculture and Fisheries Institute

Decapoda

Decapoda	

"Decapoda" from Ernst Haeckel's *Kunstformen der Natur*, 1904	
Scientific classification	
Kingdom:	Animalia
Phylum:	Arthropoda
Subphylum:	Crustacea
Class:	Malacostraca
Superorder:	Eucarida
Order:	**Decapoda** Latreille, 1802
Suborders	
Dendrobranchiata Pleocyemata See text for superfamilies.	

The **decapods** or **Decapoda** (literally "ten-footed") are an order of crustaceans within the class Malacostraca, including many familiar groups, such as crayfish, crabs, lobsters, prawns and shrimp. Most decapods are scavengers. The order is estimated to contain nearly 15,000 species in around 2,700 genera, with approximately 3,300 fossil species.[] Nearly half of these species are crabs, with the shrimp (c. 3000 species) and Anomura (including hermit crabs, porcelain crabs, squat lobsters: (c. 2500 species), making up the bulk of the remainder.[] The earliest fossil decapod is the Devonian *Palaeopalaemon*.[1]

Anatomy

As the name Decapoda (from the Greek δέκα, *deca-*, "ten", and πούς / ποδός, *-pod*, "foot") implies, all decapods have ten legs. These are in the form of five pairs of thoracic appendages on the last five thoracic segments. The front three pairs function as mouthparts and are generally referred to as maxillipeds; the remainder are pereiopods. In many decapods, however, one pair of legs has enlarged pincers; the claws are called *chelae*, so those legs may be called *chelipeds*. Further appendages are found on the abdomen, with each segment capable of carrying a pair of biramous pleopods, the last of which form part of the tail fan (together with the telson) and are called uropods.

Classification

Classification within the order Decapoda depends on the structure of the gills and legs, and the way in which the larvae develop, giving rise to two suborders: Dendrobranchiata and Pleocyemata. Dendrobranchiata consists of prawns, including many species colloquially referred to as "shrimp", such as the "white shrimp", *Litopenaeus setiferus*. Pleocyemata includes the remaining groups, including "true shrimp".[2] Those groups which usually walk rather than swim (Pleocyemata, excluding Stenopodidea and Caridea) form a clade called Reptantia.[3]

The following classification to the level of superfamilies follows De Grave *et al.*[]

Order **Decapoda** Latreille, 1802

- Suborder **Dendrobranchiata** Bate, 1888
  - Penaeoidea Rafinesque, 1815
  - Sergestoidea Dana, 1852
- Suborder **Pleocyemata** Burkenroad, 1963
  - Infraorder **Stenopodidea** Bate, 1888
  - Infraorder **Caridea** Dana, 1852
    - Procaridoidea Chace & Manning, 1972
    - Galatheacaridoidea Vereshchaka, 1997
    - Pasiphaeoidea Dana, 1852
    - Oplophoroidea Dana, 1852
    - Atyoidea De Haan, 1849
    - Bresilioidea Calman, 1896
    - Nematocarcinoidea Smith, 1884
    - Psalidopodoidea Wood-Mason, 1874
    - Stylodactyloidea Bate, 1888
    - Campylonotoidea Sollaud, 1913
    - Palaemonoidea Rafinesque, 1815
    - Alpheoidea Rafinesque, 1815
    - Processoidea Ortmann, 1896
    - Pandaloidea Haworth, 1825
    - Physetocaridoidea Chace, 1940
    - Crangonoidea Haworth, 1825
  - Infraorder **Astacidea** Latreille, 1802
    - Enoplometopoidea de Saint Laurent, 1988
    - Nephropoidea Dana, 1852
    - Astacoidea Latreille, 1802
    - Parastacoidea Huxley, 1879
  - Infraorder **Glypheidea** Winckler, 1882
    - Glypheoidea Winckler, 1882
  - Infraorder **Axiidea** de Saint Laurent, 1979b
  - Infraorder **Gebiidea** de Saint Laurent, 1979
  - Infraorder **Achelata** Scholtz & Richter, 1995
  - Infraorder **Polychelida** Scholtz & Richter, 1995
  - Infraorder **Anomura** MacLeay, 1838

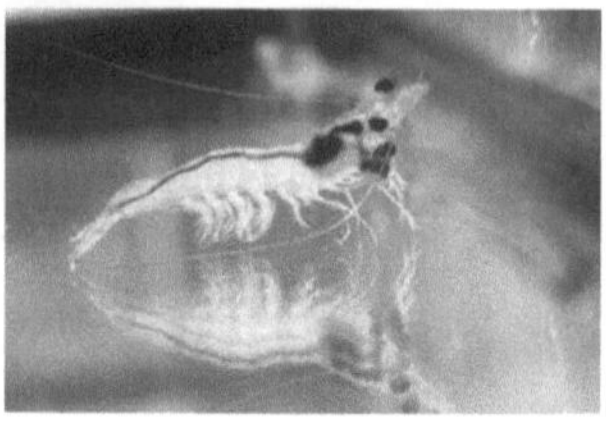

Whiteleg shrimp, *Litopenaeus vannamei* (Dendrobranchiata: Penaeoidea)

Heterocarpus ensifer (Caridea: Pandaloidea)

Austropotamobius pallipes (Astacidea: Astacoidea)

- Aegloidea Dana, 1852
- Galatheoidea Samouelle, 1819
- Hippoidea Latreille, 1825a
- Kiwaoidea Macpherson, Jones & Segonzac, 2005
- Lithodoidea Samouelle, 1819
- Lomisoidea Bouvier, 1895
- Paguroidea Latreille, 1802

- Infraorder **Brachyura** Linnaeus, 1758
  - Section Dromiacea De Haan, 1833
    - Dromioidea De Haan, 1833
    - Homolodromioidea Alcock, 1900
    - Homoloidea De Haan, 1839
  - Section Raninoida De Haan, 1839
  - Section Cyclodorippoida Ortmann, 1892
  - Section Eubrachyura de Saint Laurent, 1980
    - Subsection Heterotremata Guinot, 1977
      - Aethroidea Dana, 1851
      - Bellioidea Dana, 1852
      - Bythograeoidea Williams, 1980
      - Calappoidea De Haan, 1833
      - Cancroidea Latreille, 1802
      - Carpilioidea Ortmann, 1893
      - Cheiragonoidea Ortmann, 1893
      - Corystoidea Samouelle, 1819
      - Dairoidea Serène, 1965
      - Dorippoidea MacLeay, 1838
      - Eriphioidea MacLeay, 1838
      - Gecarcinucoidea Rathbun, 1904
      - Goneplacoidea MacLeay, 1838
      - Hexapodoidea Miers, 1886
      - Leucosioidea Samouelle, 1819
      - Majoidea Samouelle, 1819
      - Orithyioidea Dana, 1852c
      - Palicoidea Bouvier, 1898
      - Parthenopoidea MacLeay,
      - Pilumnoidea Samouelle, 1819
      - Portunoidea Rafinesque, 1815
      - Potamoidea Ortmann, 1896
      - Pseudothelphusoidea Ortmann, 1893
      - Pseudozioidea Alcock, 1898
      - Retroplumoidea Gill, 1894
      - Trapezioidea Miers, 1886
      - Trichodactyloidea H. Milne-Edwards, 1853
      - Xanthoidea MacLeay, 1838
    - Subsection Thoracotremata Guinot, 1977

Upogebia deltaura (Gebiidea: Upogebiidae)

California spiny lobster, *Panulirus interruptus* (Achelata: Palinuridae)

Polycheles sculptus (Polychelida: Polychelidae)

Australian land hermit crab, *Coenobita variabilis* (Anomura: Paguroidea)

- Cryptochiroidea Paul'son, 1875
- Grapsoidea MacLeay, 1838
- Ocypodoidea Rafinesque, 1815
- Pinnotheroidea De Haan, 1833

Atlantic blue crab, *Callinectes sapidus* (Brachyura: Portunoidea)

See also

- List of Atlantic decapod species
- Phylogeny of Malacostraca

References

External links

- Decapod Crustacea (http://decapoda.nhm.org/) "Tree of Life" page at the Natural History Museum of Los Angeles County
- Decapoda (http://www.dmoz.org/Science/Biology/Flora_and_Fauna/Animalia/Arthropoda/Crustacea/Malacostraca/Decapoda//) at the Open Directory Project

Article Sources and Contributors

Hepatoporus *Source*: http://en.wikipedia.org/w/index.php?title=Hepatoporus *Contributors*: NielsenGW, Stemonitis

Crab *Source*: http://en.wikipedia.org/w/index.php?title=Crab *Contributors*: 12bsilling, 12chancy, 2001:4930:58:0:1293:E9FF:FE5F:6599, 21655, 2602:306:CD94:8AF0:1100:E6B5:2798:FAA5, 5tgoodyer, 83d40m, A. Parrot, ABF, AJR, AJashley, Aadam786, Aaron Hurst, Aaron Schulz, Academic Challenger, Achowat, Acroterion, Acsenray, Adambro, Adamslattery54, Addshore, Adityamadhav83, Aeonimitz, AgainErick, Ahoerstemeier, Aitias, Aksi great, Alansohn, Aledubr, AlexCovarrubias, AlexPlank, AlexiusHoratius, Alicialovesanna17, Alleison123, AnakngAraw, Anetode, Angela, Anna Frodesiak, Anonymi, Antandrus, Antonio Lopez, Apparition11, Argo Navis, Arsenalchampsleague, Artrush, Aruton, AshLin, Ashkânián, Astral, AstroHurricane001, Atadrus, Avoided, BLUNTBILL, Baa, Bacchus87, Badsoccerplayer1, Barneca, BarretB, Beeblebrox, Bender235, Bennybp, Bentz16, Biff33, Bita ali, Blackngold29, Blanchardb, BlondieUK1988, Bobo192, Bogey97, Bongwarrior, Bornhj, Bradeos Graphon, Brian Crawford, Brothejr, Bueford243, Bwsmith1, C628, CTZMSC3, CWY2190, Cactus.man, Cadwaladr, Caiaffa, Cakebiscut112, Caltas, Calvin 1998, CanadianLinuxUser, Cape cod naturalist, CapitalR, Capricorn42, Carmichael, Catgut, Cbonsall, Ceiteag, CelestialSpore, Cenedlaetholwr Cymreig, Chmee2, Chris 73, Chris Mason, Christian75, Chun-hian, Clarince63, ClockworkLunch, Cmichael, Cnilep, Cold learner, Contagious foot, Coralnewby, Cork Crab, Cpl Syx, Cquan, Crab1111, CrabbMaster, Cremepuff222, Crustaceanguy, Cyberevil, DARTH SIDIOUS 2, DMacks, DRTllbrg, DVdm, Dale 79, Damicatz, Dandaman69dan, DanielDeibler, Danogo, Darkwind, Darry2385, Darth Panda, David efc, DavidMcCabe, DawnisuponUS, Dawson, DeadEyeArrow, Decltype, Defender Of Justice, Delirium, Denisarona, DerHexer, Dhp1080, Diannaa, Dieb4ulive, Discospinster, DivineAlpha, Dj Capricorn, Djheini, Dlohcierekim, Doctoroxenbriery, Doodle-he-like, Dr Greg, Dr.frog, DrChrissy, DuncanHill, Dycedarg, Dysepsion, Dysmorodrepanis, DéRahier, E Wing, E. Ripley, EOBeav, ERcheck, EdHarris, EddEdmondson, ElationAviation, Elipongo, Elkman, Elvishonkey, EngineerScotty, Enigmaman, Enlil Ninlil, Enomines, Enviroboy, EoGuy, Epipelagic, Equazcion, Eric-Wester, EricWesBrown, Ericb100, Erik9, Erikisabadass, Erpert, Esculapio, Esoxid, Essjay, Estevoaei, Eugene van der Pijll, Everyking, Excirial, Falcon8765, Falcon9272, Faradayplank, Farosdaughter, Favonian, Fergbayn, Fieldday-sunday, Fifthman, Flyfredred2010, Food64, Frankie0607, Frankwalko567, Frecklefoot, Fsotrain09, FuegoFish, FunPika, Fuzheado, Fvasconcellos, Gadfium, Gaius Cornelius, Gawaxay, Gdr, Gentgeen, Gikü, Ginsuloft, Glacialfox, Glane23, GlassCobra, Glenn, Gogmandley, Gogo Dodo, GorillaWarfare, Gouerouz, Gpietsch, GraemeL, GrahamBould, Greasymen, Gritironskillet, Ground Zero, Gtstricky, Gunmetal Angel, Gurch, Gveret Tered, Hadal, HaeB, Haham hanuka, HarlandQPitt, Haukurth, Hellz the smarter, HenryXVII, Hersfold, HexaChord, Hi675, HobosTakeMyCash, Hohum, Howcheng, Husond, Hut 8.5, Hydrogen Iodide, II MusLiM HyBRiD II, IRP, Ianeiloart, Igoldste, Imaninjapirate, ImperatorExercitus, Imperial Star Destroyer, Instinct, Invertzoo, Iohannes Animosus, Ioscius, Iridescent, Iroony, Irwangatot, ItsZippy, Ixfd64, J Milburn, J.delanoy, JBsupreme, JForget, JK the unwise, JNW, Jab843, Jackfork, Jacksinterweb, Jagun, Jakecp1, Janderk, Janeb92, Jauhienij, Javert, Jcf0987, Jcw69, Jeff Muscato, Jeffrey Mall, Jellybean308, Jerrymai1995, Jhinman, Jim1138, Jimmyflimingo, JoJan, John254, Jojhutton, Jonadin93, Jpjazz, Ju9ijjhibjuyhbo, JuJube, Juliaisawesome, Juliancolton, JuneGloom07, KHM03, Kaarel, Kakofonous, Kamokilla, Kandar, Karbear0512, Katalaveno, Katherine, Keilana, Khukri, KiernanBP1, Kineticman, King Zebu, King fisher322, Kingpin13, Kitch, Kjell Knudde, Kl1314, Klkuo, KnightLago, KnowledgeOfSelf, Koona12, Kooplatroo, Kory4009, Kurowoofwoof111, Kuru, Kî-tuk-thù, L Kensington, LAX, Labnoor, Ldhippo, Le Scarlet Douche, LeFapOnymous, LeaveSleaves, Lectonar, LemonTwinkle, LeonardNimoy, Lewa of the wind, Lexor, Lfstevens, Lightmouse, Ligulem-s, Lilac Soul, Lippett, Little Mountain 5, Lkiller123, Lommer, Longbowman, Lopo, Lord Cancer, Luna Santin, Luvs2spooge, Lycaon, Lyndsayruell, MBisanz, MC10, MER-C, MJ94, MPerel, MZMcBride, Macedonian, Macy, Macy emerald, Madhero88, Magister Mathematicae, MagnificentMafiaMan, Magog the Ogre, Mailer diablo, Majorly, Malagua, Manberg, Marek69, Mark Arsten, MarsRover, Martarius, Martijn Hoekstra, MartinUK, Masonislimpy, MasterChief117orignal, Matdrodes, Materialscientist, Mathew5000, Matthew21995, Mattisse, Maxí, Mbz1, Mdebets, Mdp017, Meaghan, Meegul304, Mejor Los Indios, Menchi, Mentisock, Mesh1ka, Metricopolus, Mexcellent, Mgarcia001, Mgiganteus1, Michaelas10, Michaelb304, Mindmaker, Mitchgb123, Moewwe54, Montrealais, Morganism07, Mr Stephen, Mrown1, Mrt3366, Mumbo jiggy, Mwng, Nauticashades, NawlinWiki, Nbarth, Ncsaint, Neelix, Nekonoko, Nemememe99, Nhobgood, Nihiltres, Nimbs123, No Guru, Nomorenonotnever, Ocaasi, Olly150, Omgidontno, Omicronpersei8, Omulazimoglu, Onebravemonkey, OpenMind, Orangemarlin, Orion Minor, Otolemur crassicaudatus, OwenX, Oxymoron83, P. S. F. Freitas, Pakshya, Patstuart, Paul Ferber, Pearle, Pedro, PereiraALL81, Peter Isotalo, Peter.C, PeterCanthropus, Pharaoh of the Wizards, Philip Trueman, Philosophistry, Piano non troppo, Piisamson, Piledhigheranddeeper, Pinethicket, Pip2andahalf, Plesiosaur, Polluxian, PookeyMaster, Portillo, Pparazorback, Prashanthns, Prys2010, Puchiko, Pyfan, Qaddosh, Qweef, R'n'B, Radical Mallard, RadioGirlDone, Radon210, RainyTeacher, RapidR, RaseaC, RazorICE, Rcarsey, Redvers, RexNL, Rhyshumphrey6, Rich Farmbrough, Rileykiller5, Rlwilliams34, Rockoval, RogerShermanIsBoss, Ron Ritzman, Ronhjones, Rooptedoop, Rrburke, Rulerofsoccer, Runt123, Rushbugled13, Ryan, Ryan Vesey, S9ssaffe1, STRONGlk7, Saladfingers4life, SchfiftyThree, SchuminWeb, ScottSteiner, Seaphoto, Seb az86556, Seba5618, Seglea, Sendbinti, Several Pending, ShadowKinght, Shafiru, Shaggycats, Shinnin, Shirulashem, Shoessss, Shrine of Fire, SidP, Singularity42, Sintaku, Sir sexah, Sisroxtexas, SixTimes7, Sj14786, Sjakkalle, Sjb0926, Skater, Skew-t, Sky Attacker, SkyWalker, Skyezx, Smallweed, Smolamj, Snow Blizzard, Snowolf, Soapthgr8, Soccerplayer75, Sokeater, Some jerk on the Internet, Sophdan, Soup man, SparrowsWing, Spaspaz, Spaully, Speed0012, Spencer, Spiegal, SpyMagician, Srgazuzu, Staffwaterboy, StaticGull, Stefan, Stemonitis, Stephenb, Stevefis, Steven Zhang, Stiklu, Stwsoad, Suffusion of Yellow, Summer Killian, Sunderland06, Sup889, Surachit, Susannelauro, Susfele, Sylvsie35, TANAKA Juuyoh, Tabletop, Tarinth, Tastehchickenwing, Tastehchickenwingreturns, Tbhotch, TeaDrinker, Tedmund, TehUltimateGuyOfUltimatePwnage, Tehinteractive, Terromamer7, Tewy, Texhausballa, Tgeairn, Tgmonkeyman, Thatguy1010, The Anonymouse, The Haunted Angel, The High Fin Sperm Whale, The Masked Turnip, The Thing That Should Not Be, The Utahraptor, The sharp wolf, TheGrimReaper NS, TheNewPhobia, TheRanger, Therazor35, Thompson.matthew, Tiddly Tom, Tide rolls, TimVickers, Tins128, Tmorgan147, Tombomp, Tomcat, Tommy2010, Travelbird, Tresiden, Trhazuka92, TrulyBlue, Trusilver, Tuspm, Tylershannon123, Ufim, Ukexpat, Ulric1313, Uncle Dick, Uncle Milty, UncleTroll, Unschool, Usb10, Uspn, Ute in DC, VGer!, Valermos, Vanished user ois9h3jfujw4f, Vanished user uih38riiw4hjlsd, Versus22, Vezhlys, Vicenarian, Violetriga, Viriditas, Visik, Voldemar69, Vrenator, WAvegetarian, WadeSimMiser, WarthogDemon, Wayne Slam, Webclient101, Wenli, WereSpielChequers, Wfaulk, Wiki alf, Wikipelli, Wild.error, William Avery, William347, WilliamH, Wimt, Wine Guy, WolfmanSF, Wuhwuzdat, Wumb0man9000, Wōdenhelm, Xaosflux, Xavier212191, Yintan, Zachzolty, Zapf123, Zaqarbal, ZhiRandom, ZooFari, Zxzxzxg, 1220 anonymous edits

Xanthidae *Source*: http://en.wikipedia.org/w/index.php?title=Xanthidae *Contributors*: 80.255, AvicAWB, CultureDrone, Dawson, Droll, Dysmorodrepanis, Gdr, GrahamBould, Isfisk, Obsidian Soul, Skapur, Stemonitis, TheParanoidOne, Updatehelper, 9 anonymous edits

Animal *Source*: http://en.wikipedia.org/w/index.php?title=Animal *Contributors*: (jarbarf), -Paul-, 123forever456, 1313 South Harbor Blvd. Anaheim, CA, 14Ave, 168..., 2004-12-29T22:45Z, 28421u2232nfenfcenc, 2D, 4175shelton, 505, 545lljkr, 9258fahsflkh917fas, A-giau, Aaadddaaammm, Aaron Schulz, Aassaas, Abductive, Abigail-II, Abyssal, Academic Challenger, Acalamari, Acdx, Ace ETP, Acebulf, Achiwiki356, Adam78, Adambro, Adashiel, AdjustShift, Adriano1995, Adrigon, Aeonx, Aethon8, Aff123a, Afgheys12, AgainErick, Ahambhavami, Ahoerstemeier, Aidarzver, Aim Here, Airoman, Akendall, Akral, Aksi great, Alainacmitchell, AlanH, Alansohn, Alaska great, Albatross Loves You, Ale jrb, Aleenf1, Alex croff, Alexandria, AlexiusHoratius, Alexy527, Ali, Aliasxerog, Alimon, AllGloryToTheHypnotoad, Ally loves softball, Alterego, Alvin-cs, Amaltheus, Amazins490, Anastrophe, Andjosaus, Andre Engels, Andrewrp, Andux, Andycjp, Angela, Angr, Animum, Ann Stouter, Anna Frodesiak, Anna Lincoln, Anonymous Dissident, Antandrus, Anthere, Anthony1001294, Anthony717, AnthonyQBachler, Anxietycello, Apostrophe, Apparition11, Aquadiamond3, Aquamarine04, Arakunem, Arcarius, Archer3, Arisa, Armaghanje, Art LaPella, Arthur Rubin, Ashmoo, Astropithicus, Aswesee, Atchernev, Athaler, AtikuX, AtonX, Atropos, AuburnPilot, Aude, AutomaticWriting, Avenged Eightfold, Avoided, AxelBoldt, Ayrton Prost, AzaToth, Azhyd, BD2412, Baccyak4H, Ballista, Bananarofl, Barbary lion, Barneca, Barrylb, Baseballfreakrc, Basharh, Bassbill, Bassbonerocks, Bbatsell, Bbvslg, Bcasterline, Becie01, Beeblebrox, BeefRendang, Before My Ken, Belligero, Ben-Zin, Ben.c.roberts, BenJWoodcroft, Bendzh, Benjamint444, Benjiboi, Bensaccount, Beta m, Bgold, Bidiot, Big Bird, BigChicken, Bigbossfarin, Bigwang6969, Billy775, Biology999, Blanchardb, Blarg3001, Blarrrgy, BlastOButter42, Blaxthos, Blehfu, Bobbean, Bobisbob2, Bobo192, Bobthebuilderbuildsagain, Boccobrock, Boch501, Bomac, BonesBrigade, Bongwarrior, Bookermorgan, Booyabazooka, BorgQueen, Boston, Brambleclawx, Branddobbe, Brian0918, Brianga, Briséis, BrokenSegue, Brossow, Brownsc, BryBla, Bschrader, BuckeyeMoe, Bumbbumb, Bumholio, Bunchofgrapes, Burmeister, CALR, CDepiereux, CJLL Wright, Caesura, Caltas, Can't sleep, clown will eat me, CanadianCaesar, Canderson7, CanisRufus, Capnquackenbush, Carabinieri, CardinalDan, Carl.bunderson, Carlobus, Casper2k3, Causa sui, Cdc, Cdernings, Cekli829, Celithemis, Cephal-odd, Chamal N, Champlax, CharlesHBennett, Chaser, Cherry blossom tree, Chipmunkdavis, Chl, Chmod007, Choanoflagellate, Chocolaterain69lolrofl, Chowchowpoop, Chris j wood, Chris the speller, Chrisd87, Chrislk02, Christian List, Christopher Higgins 93, Chuunen Baka, Ciaccona, Cjc0013, Cleared as filed, Cliff smith, ClockworkSoul, Closedmouth, Cloudy fox 001, Cmorales13990, Cod6user, Cody12343, Coelacan, CohenTheBavarian, Colin Kimbrell, Collieman, CommonsDelinker, Conorsampson, Conti, Conversion script, Cool3, Corncake23, Corpus juris, Corpx, Corvus coronoides, Cory Crowley, Countney, Courcelles, Cow monkey111, Cp111, Cpl Syx, Creeing, Crip killah, Crum375, Crusadeonilliteracy, Crustaceanguy, Crzrussian, Csparr, CupOfRoses, Cupcakelover101, Curps, Cyborg1981, Cyclopia, Cyrius, D1sn1gg4, DARTH SIDIOUS 2, DVD R W, Dada43, DagnyB, Dal3164, Dampir, Danger, Daniel C. Boyer, DanielCD, Dar book, DarkFalls, Darkclothes violentpose, Darkmiles22, Darkwind, Darolew, Darth Chyrsaor, Darthgriz98, Dasani, Davewild, David D., David Munch, David0811, Davidlittle9, Dawn Bard, Dcandeto, DeLarge, DeadEyeArrow, Deeptrivia, Delanaschroeder, Delbert Grady, Delirium, Delldot, Demmy, Dendodge, Deor, DerHexer, Dforest, Dgw, Diagonalfish, Diannaa, DifiCa, Dinno, Dinoguy2, Dinokid, Dipper32, Disavian, Discospinster, Dissimul, DivineAlpha, Dj55, Dman64, Doc glasgow, Dockcharlotte, Dodo bird, Dojarca, Dolpihnfan123, Donald Albury, Donarreiskoffer, DonnaSG, Dorvaq, Doug Bell, DrMicro, Dracka, Drat, Dratman, Drc79, Dreaded Walrus, Dreadstar, Drew R. Smith, Drmegabite, Duanedonecker, Dude1818, Dudemeister1234, Dustimagic, Dysepsion, Dysmorodrepanis, Dysprosia, E2eamon, EM85462, EPM, ESkog, EWS23, EamonnPKeane, Earlypsychosis, Earth'sbuddy, EarthPerson, East of Borschov, Ebeisher, Ed Poor, Ed g2s, Edk, Edom23, Eightofnine, EikwaR, El C, ElLoEvErYbOdY, ElePANTS, Eleassar, Elias1richa, Eliezg, Eliz81, Elmo Zat, Eluchil404, Eminem95941, Emperorbma, EncMstr, EncycloPetey, Endangered11, Enigmaman, Enviroboy, Epbr123, Eran, Erick7723, Erik9, Esanchez7587, Escape Orbit, EscapingLife, Esn, Esurnir, Eu.stefan, Eugene van der Pijll, Everyking, Excirial, Exert, Eyu100, FF2010, Fabiform, Facts707, Fadesga, Fang 23, Farrahi, Farsight001, Feedloadr, Felizdenovo, Femto, Ferel destroyer, Fgr1994, Fieldday-sunday, Fil21, Filelakeshoe, Finlay McWalter, FireOfTheNight, Firsfron, FisherQueen, Flat turds, Flewis, Flo98, Flowerpotman, Foosher, Francisco Valverde, Frap, Freakofnurture, Fredvanner, FreplySpang, Fritoflyer, Frubey, Fullz, Furkaocean, Furkhaocean, Fvasconcellos, Fvw, Fæ, GB fan, GTPoompt, Gaaras-brother, Gabbe, Gadfium, Gaff, Gaius Cornelius, Gakrivas, Galoubet, Gamesmasterg9, Ganesh Paudel, Gavrilis, Gcjdavid, Gdr, Geni, GeorgeMoney, Gfoley4, Gianfranco, Giftlite, Gilliam, Gimmetrow, Gingerman1234567, Ginkgo100, Gjd001, Glenn, Gnostrat, Go for it!, Gogo Dodo, Goldom, Goodnightmush, Gothicrocka, GraemeL, Graham87, GrahamBould, Graminophile, Grblundell, Greeener, Greswik, Grundle2600, Gscshoyru, Guanaco, Gundato, Gurch, Gutworth, Gwernol, Gypsy Trash, H3110kitty, Hadal, Haham hanuka, HalJor, HalfShadow, Halfcrowd123, Halo master chief, Halosaver, Ham1999, HappyBlades, HappyKarMar, Happyface0000, HarlandQPitt, Hbomb194, Heebiejeebieclub, Heegoop, Heimstern, Helicoptor, Hellbus, Hello32020, Heman, Hephaestos, Hermanjr, Herr Gruber, Hibana, Hihi582, Hordaland, Hotie1234, Hu12, HughesJohn, HunterAmor, Hut 8.5, Hyebeh, I dream of horses, IAMTHEEGGMAN, ICAPTCHA, II MusLiM HyBRiD II, Ianml, IdahoBoy, Idont Havaname, Ifnord, Indoles, Iner22, Insanity Incarnate, Intelligentsium, Into The Fray, Irbisgreif, Iridescent, Ixfd64, J. Spencer, J.delanoy, JAKLAM, JForget, JIP, JLaTondre, JNW, JTSomers, JULIA RADESCU IS A SKIVER!, JYolkowski, JamesAM, Jamesofur, Jamsamree, Janderk, Jannex, Jarble, Jason Quinn, Jasongetsdown, Jauerback, Javert, Jawsx188, Jay, Jay2332, JayW, Jaymckenzie, JedOs, Jeendan, Jefffire, Jeffrey Mall, Jeffrey O. Gustafson, Jeremy Visser, Jeremy-list, Jerkov, Jeronimo,

Jesseblum, Jewman55, Jfhutson, Jibbo Whales, Jiffijosh, Jimp, Jj137, Jklin, Jmv2009, Jnelson09, Jnothman, JoJan, Joeler, Joelr31, John Price, JohnOwens, Johnleemk, Joka, JonasRH, Jordanlou, Jordy321, JorisvS, Jose77, Josh Grosse, Joshnadler, JoshuaZ, Jossi, Joyous!, Jpgordon, Jrockley, Juansidious, JudelFoir, Jules.LT, Julesd, Juliancolton, Juniuswikia, Jusdafax, Justanothervisitor, Justin, JustinHagstrom, Justsomecoolguyhomie, Jwanders, Jwray, Kablammo, Kaisershatner, Karlhahn, Kateyritchie, Kbh3rd, Keilana, Kevin B12, Kf4bdy, Khoikhoi, Kig108222, KillerChihuahua, King of Hearts, Kingturtle, Kintetsubuffalo, Kirill Lokshin, Kittystarkails, Knowledge Seeker, KnowledgeOfSelf, Knowledgeum, Koavf, Koolpapajohn, Korg, Kosigrim, Kostya, Kpjas, Krawi, Krn10919, Ktaylor, Kumioko (renamed), Kurt Shaped Box, Kuru, Kvamme, Kwamikagami, Lachatdelarue, Laguna117, LarryQ, Launchballer, Lcarscad, LeadMeOn, Leadwind, LeaveSleaves, Lectonar, LedgendGamer, Lee Daniel Crocker, Legions pker, Leif, Len124, Leonidasthespartan, LessHeard vanU, LibLord, Lightmouse, Lights, Linmhall, LiquidGhoul, Lissisa2003, Littlec911, Llamaloop2056, Llull, Lomn, Looie496, Lord Patrick, Lord Pistachio, Lost tourist, Lotusduck, Lovely 101, Loving666, Luk, Lupin, Lupo, Lycurgus1920, Lygor, MER-C, MK8, MKoltnow, MPF, MZMcBride, Mad Greg, Madpassion, Maegara, Magister Mathematicae, Mahbabygirl, Mahdi1ray, Majorly, Malljaja, Malo, Man vyi, Mandyjb, Manuel Anastácio, Mapocathy, Marbehraglaim, Marcus Qwertyus, MarkSutton, MarsRover, Marskell, Martin451, Materialscientist, Mattdaman72, Matthew Wilson, Matthew Woodcraft, Matthew Yeager, Matticus78, Mav, Maxmuscroft, Maxzc, Meaningful Username, Measure, Medeis, Meemlo, Mego'brien, Melsaran, Meno25, Menswear, Merope, Merqurial, MetsFan76, Mgiganteus1, Michaelfavor, MihaS, Mike Rosoft, Mike V, MikeLynch, Miklo68, MiltonT, Mindmatrix, Mindofcameron, Miquonranger03, MisfitToys, Missmarple, Mitternacht90, Mlb32, Molinari, Moon&Nature, Moondyne, Mowgli, Mperkins, Mpg1995, Mr Stephen, Mr.TimTam, MrOllie, Mschel, Mspraveen, Muhaidib, Munita Prasad, Mushroom, Mwanner, Mweites, MySummerJob, Myanw, Myok, Myriologist, Mywebsearch, N5iln, Nanobug, Narayanese, Nathanww, NatureA16, Navid7, NawlinWiki, Nbarth, Ndenison, Neelix, Nehrams2020, Neil916, NeoJustin, Neoprote, NepGrower, Netoholic, Neurolysis, NewEnglandYankee, Newnoise, Newportm, Nick Number, Nickshanks, Nicolharper, Nifky?, Nihiltres, Nikai, Nikkimaria, Ningauble, Ninjatacoshell, Nivix, Nm4321, Noleander, NorCal764, Norm, NotWith, Notedgrant, NuclearWarfare, Nuttycoconut, Nwbeeson, OMGsplosion, ONUnicorn, Obsidian Soul, Ocatecir, Oda Mari, Ohconfucius, Ohnoitsjamie, Old Moonraker, Oli Filth, Omicronpersei8, Onco p53, Oneirist, Onorem, OwenX, Oxymoron83, Ozgod, PBP, PGWG, Pabbles, Paracel63, Parker97558, Paroxysm, Paskari, Pass a Method, Patrick, Paul 012, Paul August, Paul Erik, Paul Murray, Pax:Vobiscum, Pbroks13, Pcb21, Pearle, Pekinensis, Pelister, Penwhale, Pepper, Persian Poet Gal, Pethr, Petrb, Pgk, Phantom784, Phantomsteve, Phen777, Phenz, Philip Trueman, PhilipO, Philx9771, PhnomPencil, Phoebepheebs, Phpham31, PiMaster3, Piano non troppo, Picus viridis, PierreAbbat, Piet Delport, Pigman, Pikachu26, Pilotguy, Pinkadelica, Pippu d'Angelo, PizzaBox, Plasynins, Plugwash, Poetaris, Poindexter Propellerhead, Porterjoh, Postglock, Prari, PrestonH, Pretty Green, Probatio Pennae, Prodego, Programingcraze007, Programmer13, Prolog, Psychonaut, Ptcamn, Pthag, Punaboy7, Punarbhava, Qazwax123, Quadell, Quadrius, Quahog5News, Quaque, QuartierLatin1968, Qwertzy2, Qxz, R9tgokunks, RBJ, Raamah, Randomuselesstrivia, Randy Johnston, Ranemanoj, Ranveig, Rasmus Faber, Raven4x4x, Razao131, RazorICE, Rdsmith4, Red Director, Red Winged Duck, RedHillian, RedWolf, Redneck017, Reformedman, Rellis1067, Res2216firestar, Rettetast, Reuvenk, Revolver, RexNL, Reyk, Reywas92, Rich Farmbrough, Richard D. LeCour, Richard001, RichardF, Rickterp, Rjwilmsi, Robchurch, Robert is da man, Robin Williams, Robinh, Robomaeyhem, RockMFR, Rockpocket, Roland Deschain, Romanm, Roonyroon, Roozie12, Rory096, Rose Garden, RossPatterson, RoyBoy, RoySmith, Rrburke, Rror, Rsrikanth05, Rutalkin2me0574, Ruud Koot, Ryan, Ryan Postlethwaite, Ryansca, SCS100, SKUK98, SMC, ST47, Sacredseal, Sainsf, Sakana-rin, Samak47, Samirwin, Sampi, Samsara, Samw, SandyGeorgia, Sannse, Saperaud, Sarcasticmac, Sarpasarpeti101, Sarranduin, Saturn star, Savant13, Scarypi33, SchfiftyThree, Schmei, Schzmo, Science1563, Scientizzle, Scohoust, Scorpiuss, Scott Ritchie, Scottalter, Sdornan, SeanLegassick, Seanutbutter, Seb az86556, SebastianHelm, Semperf, Senator Palpatine, Sengkang, Sentience, Serpent's Choice, Sethdeperot, Seven7, ShakingSpirit, Shamesspwns, Shanes, Shatosy, Sheep2000, Shindo9Hikaru, Shmget, Shoaler, Shoessss, SilentBobxy2, SiliconDioxide, Simeon H, SimonP, Sjakkalle, Skippy5490, SkyBoxx, Skysmith, Slowking Man, Smack, SmartGuy Old, Smb3guy, Smeazel, Smerdis, SmilesALot, Smith609, Smithy01, Smohammed2, Snarfsnarf, Snowmanradio, Soccerguy7735, Soliloquial, Somecrazybastard13, SouthernNights, Sp, Sp3000, SpaceFlight89, SparrowsWing, Spe0059, SpeedyGonsales, Spencer, Spiritualism, Splash, Spotty11222, SpuriousQ, Spyderxskierx4, Squash, Squids and Chips, Srikeit, Staffelde, Stan Shebs, Starexplorer, Starnestommy, Stefan, Stefan Udrea, SteinbDJ, Steinbach, Stemonitis, Stephen MUFC, Stephen52, Stephenb, SteveRamone, Steven Walling, Steven407653, StevenGantzer, Stevertigo, Stewartadcock, Stfg, Storm Rider, Strongbadman, Super678, Susan Davis, SusanLesch, Sweetness46, Sweetteen1, Syrthiss, TEB728, TMC1221, TUF-KAT, TWCarlson, Tablelajk3, Tableofwood, Tailsdoll666, Tannin, Tapir Terrific, Tau'olunga, Taybabe, Tbhotch, Tchams04, Telephone123, Tellyaddict, Template namespace initialisation script, Tero, TerryNF, Tgeairn, ThaddeusB, The Anome, The Giant Puffin, The High Fin Sperm Whale, The Missing Hour, The Rambling Man, The Thing That Should Not Be, The guy who roolz, The mysterious gypsy, The undertow, TheDisser, TheLimbicOne, TheProject, TheThingy, TheaterMarine, Thejerm, Thysko, Tide rolls, Tigerhawkvok, TimVickers, Timrem, Tinytipper, Tiptoety, Titoxd, Tom harrison, Tom murray rust, Tomashcu60, Tomi, Tompsett, TonyBallioni, Tox, Tpbradbury, Transmogribenno, Transource, Traroth, Treatie, Treehouse372, Tregoweth, Triddle, Trilobite, TripleAven, Troge20, Troogleplex, Trysha, Tungsten, TutterMouse, Twintone, Typedia, Uavbuff, Ucucha, Ukt-zero, Ulric1313, Untitledmind72, Unyoyega, Urocyon, UserDoe, UtherSRG, Uyvsdi, Vahid11, Vaishu2, Vandal B, Vanished User 4517, Vanished user 38hfun34tunkewfij4t, Vanished user uih38riiw4hjlsd, Vasudhathegreat, Veinor, Vicki Rosenzweig, Vig vimarsh, Vincentv85, Violetriga, Virek, Voortle, Vrenator, Vsmith, Vyyvg, WJBscribe, WODUP, Wack'd, WadeSimMiser, Waggers, Wajidkanju, Walter 1080, Watcharakorn, Wavelength, WawaNJ, Welcome7ones, WereSpielChequers, Werothegreat, Wetman, Whiskey in the Jar, Why Not A Duck, Wiki alf, WikiCantona, WikiExpertEditor, Wikidudeman, Wikilixo, Wikiman13579, Wikiniki, Wildhartlivie, Wildnox, Will Beback Auto, William Avery, Williamb, Willking1979, Wimt, Wirbelwind, Wireless Keyboard, Wknight94, Woer$, Wonderland42, Woohookitty, WormRunner, Wragge, Wtmitchell, Wysprgr2005, XJaM, Xanchester, Xerxesnine, Xev lexx, Xiaphias, Ximsuchastudx, Xwu, Xxxzenicxxx, Yamamoto Ichiro, Yaninass2, Yanksox, Yansa, Yekrats, Yetisyny, Yidisheryid, Yodude77, Yogiudo, Z.E.R.O., Zarcadia, Zenap, Zephyrus67, Zfr, Zgomez99, ZimZalaBim, Zoicon5, Zygomorph, Zylon, Zzuuzz, Ævar Arnfjörð Bjarmason, Ψ-113μ, 2120 anonymous edits

Arthropod *Source*: http://en.wikipedia.org/w/index.php?title=Arthropod *Contributors*: 130.94.122.xxx, 2112 rush, 213.253.39.xxx, 28421u2232nfenfcenc, 600 pound, A little insignificant, Abhishek727, Abrech, Access Denied, Ace ETP, Addshore, Adrian12695, Aiken drum, Aksi great, Alansohn, Alexei Kouprianov, Alexrudd, Algirdas, Allstarecho, AlmightyFjord, Alnagov, Amaltheus, Amezcackle, Anaxial, Andonic, Andre Engels, AndrewHowse, Andrewa, Andycjp, Angela, Ani Gublu, Anlace, Anomalocaris, AnonGuy, Antandrus, Antonio Lopez, Antzervos, Apokryltaros, Aramgutang, Arcadian, Argo Navis, Arthena, AshLin, Astroview120mm, Autodidactyl, Avoided, Awesome 210, AxelBoldt, Az1568, Azhyd, BLADELIGAR, BW52, Bart133, Bassbonerocks, Bbllr3431, Bcasterline, Beardless, Bill Darwin, Blackguy696969, Blacklunarsun, Bob Burkhardt, Bob the Wikipedian, Bobo192, Bobvila24747, Bogey97, Bomac, Bongwarrior, Book worm369, Bookermorgan, Boud, BritishWatcher, Brokenspirits, Brownpantz12, Bruce Marlin, Bryan Derksen, Bsroiaadn, Bucketsofg, Bugenstein, Burlairbaby, Burschik, Buthwiki, Bwatts1105, CIreland, CWii, Callumzammit, Calvin 1998, Can't sleep, clown will eat me, CanadianLinuxUser, Carbuncle, CardinalDan, Carel.jonkhout, Carmichael, Catgut, Causa sui, Cephal-odd, Cerealkiller13, Chacapito, ChadThomson, Changuno, Chaojoker, Chester Markel, Chienlit, Christian75, Christopher Parham, Chzz, Click23, Cliff smith, Closedmouth, Conversion script, Cotinis, Courcelles, Cquan, Craigkbryant, Crap & nuts, Crocadog, Crustaceanguy, Csparr, Curb Chain, Cyanolinguophile, D, DARTH SIDIOUS 2, DRUGMONSTER, DSatz, Danger, DanielCD, David Edgar, Davidkazuhiro, Dawn Bard, Delldot, Dendre, DerHexer, Deranged bulbasaur, Dfrg.msc, Dgw, Discospinster, Disneyw10, Dmn, DocWatson42, DrGatsby1962, DrSaturn, Droll, Drunken Pirate, Ducknish, Durova, Dyanega, Dysmorodrepanis, ESkog, EagleFan, Ean5533, EarthPerson, Eclecticology, Ed g2s, Ejosse1, Elb2000, Elockid, Eluchil404, Emperorbma, Enchilado, Enlil Ninlil, Enuja, EoGuy, Epbr123, Epipelagic, Epolk, Eras-mus, Eugene van der Pijll, Evil Monkey, FISHMAN1, Fabiform, Favonian, Feinoha, Fenoxielo, Fieldday-sunday, Florian Blaschke, Flyguy649, Foxj, Framed0000, Frank Warmerdam, Fred Condo, Friginator, FunPika, Fusionmix, Fuzzie, Gaia5074Q, Gaius Cornelius, Gareth Griffith-Jones, Gdarin, Gdr, Gfoley4, Giftlite, Gilgamesh, Gimmetrow, Glenn, Gogo Dodo, GoingBatty, Gordon Rugg, Grafen, Graham87, Grahbudd, Griff08, Griffinofwales, Ground Zero, Gtg204y, Gurch, Hadal, Haham hanuka, Hall Monitor, Hamiltondaniel, Haploidavey, Harland1, Harry491, Harryboyles, Headbomb, Hephaestos, Hezare, Hurricane Angel, Iammymom, Iamunknown, Ichigo20, Igetdatars, Igoldste, Immunize, Imolly98, Into The Fray, Iridescent, IronChris, Itzcuauhtli, J.delanoy, JDP90, JForget, JHolman, JNW, JaGa, Jab843, Jarble, Jared81, Jaredrkennard, Javit, Jefffire, Jeffq, Jerry.waller, Jim1138, Jimp, John Cline, JohnCD, Johnuniq, JorisvS, Josemanimala, Josh Grosse, Josh Parris, Jpvinall, Jrissman, Jujutacular, Julian Mendez, Jusdafax, Justin, KMDelvadiya, Kaldari, Kazvorpal, Keegan, Kerrick Staley, Kevin4868, Khalnath, Kidd Loris, Kils, Kimse, Kingpin13, Kipala, Kku, Kukini, Kuzaar, LOL, Lankiveil, Leafyplant, LeaveSleaves, Leeyhe, Lgcharlot, LieutenantLatvia, Ligulem-s, Lmslol, Lordmetroid, Lova Falk, Lukecutforth, M-le-mot-dit, M1ss1ontomars2k4, MER-C, MONGO, Macaddct1984, MacedonianBoy, Mad Greg, MadGuy7023, Magnus Manske, Malleus Fatuorum, Malo, Mani1, Marie Poise, MarkRose, Marshman, Materialscientist, Matijap, Matthew Fennell, Mattisse, Mav, Mavin94, Mbz1, McSly, Meaghan, Medeis, Mejor Los Indios, Menchi, Mentifisto, Mentisock, Mgiganteus1, Miborovsky, Michael Devore, Mifter, Mike Rosoft, Mortosthegodly, Moses sheperd, Mr. Wheely Guy, Mrthescholz, Muhk33, Muke, Naddy, Nashville Monkey, Nathanspyx, NatureA16, NawlinWiki, Ncu s0cc3r, Netkinetic, Netoholic, NewEnglandYankee, Newone, Nick4414, Nihiltres, Niluop, Nono64, Nooby god, Numbo3, Nwbeeson, O.Koslowski, OSU1980, Obsidian Soul, Oda Mari, Odie5533, Ohnoitsjamie, Oleg Alexandrov, Omicronpersei8, Ori, Oscarthecat, OverlordQ, OwenX, PDH, Pauli133, Paxsimius, Pcb21, Peachy235, PenguinJockey, People101, Persian Poet Gal, Peruvianllama, PervyPirate, Peter Isotalo, Peter coxhead, Philcha, Philip Trueman, Phlegm Rooster, PierreAbbat, Pikachuwashere, Pinethicket, Pippu d'Angelo, Planders, Plantsurfer, Plasticup, Pleasantville, Plumbago, Poindexter Propellerhead, Pokeman0, Ponder, Poop2000, Poopstain12, Poopy nose69, Possum, Prashanthns, Printer222, Programingcraze007, PuzzletChung, Pyfan, Quadell, Quinsareth, Quintote, Qxz, RN1970, Radical Mallard, Reach Out to the Truth, Rebel, Recognizance, RexNL, Rgoodermote, Rhynchosaur, Rich Farmbrough, RichardMills65, Richardcavell, Rickgrimes133, Rigby27, Rjwilmsi, Robin S, RoboRobo1, Rocastelo, Rocketrod1960, RockfangSemi, Rokfaith, Ross perot69, Rossami, Roy Bateman, Rschmertz, Ryryrules100, SMC, Samak47, SamuelCorcos, Sanfranman59, Saperaud, Sarefo, Scottalter, Sean.hoyland, Seaphoto, Seb az86556, Shakecandy1, Shall I Make Wikipedia A Boom Town?, Shanes, Shipmaster, Shyamal, Siim, Skapur, SkyMachine, Slugger, Smith609, Snowmanradio, Snowolf, Some jerk on the Internet, Sonjaaa, Spinningspark, Splurben, Spotty11222, Star27, Staticd, Steel, SteinbDJ, Stemonitis, Stephenb, Stevertigo, Stfg, Sundin1001, SuperMidget, Sushant gupta, Swagdawgs$, TBrandley, THEN WHO WAS PHONE?, TUF-KAT, Taggard, Tashawn, Tassedethe, Technopat, Tek022, Tellyaddict, Temporaluser, Thanatos666, The Thing That Should Not Be, The undertow, TheLimbicOne, Thecheesykid, Thehelpfulone, Thepoof2, Thingg, Thulcandra, Tide rolls, Tideflat, Tigershrike, TimVickers, Tintenfischlein, Tktktk, Tommy2010, Tomsowa, Tony Fox, Topbanana, TotoBaggins, Transity, Traxs7, Trevayne08, Trfasulo, Truth14, Tuspm, Twas Now, UBeR, Uncle Dick, Unfree, Universalcosmos, UsagiYojimbo, UtherSRG, Vandal B, Velella, Versus22, Vicki Rosenzweig, Violask81976, Vishnu2011, Vrenator, WadeSimMiser, WaltonArgus, WarBaCoN, Wavelength, Wayne Slam, Wchols, Weisbah, Welsh, Weregerbil, Widr, Wik, Williamb, Willking1979, Wilson44691, Witan, Wjl2, Wknight94, Wmahan, Wordwright, WormRunner, Wyatt915, Xaraphim, Xpinki96x, Yamamoto Ichiro, Yath, Yerpo, Yettie0711, Ykvach, \\\\\\pwarg\\\\\\, دجاس دجما دجاس, 1149 anonymous edits

Crustacean *Source*: http://en.wikipedia.org/w/index.php?title=Crustacean *Contributors*: 345Kai, 545lljkr, A-giau, Ace ETP, Aitias, Ajaxkroon, Alansohn, Aldaron, Alex.muller, Alvinstraight, Anaxial, Andreasmperu, Angela, Antandrus, Antonio Lopez, Anwar saadat, Apokryltaros, Arunsingh16, AshLin, Auric, Australianplankton, AxelBoldt, Azheng888, Baa, Barneca, Bartleby, Bballguy55, Big iron, BigPimpinBrah, Bleepmcbeep, Bob Burkhardt, Bobo192, Bongwarrior, Bookofjude, BorgQueen, Brighterorange, Bsungp1993, Calmer Waters, CanadianLinuxUser, Capricorn42, CardinalDan, Ceyockey, Cfrydj, Chris Roy, Citiboy2008, Coffee2theorems, CommonsDelinker, Cool Blue, Coralmizu, Corpx, Courcelles, Crudmongerer, Crustaceanguy, Ctbolt, CutMet, Cyanolinguophile, DVdm, Damicatz, Dan Koehl, Danielstoupin, Danny, DerHexer, Discospinster, Dlloyd, Dmn, DocWatson42, DopefishJustin, DoubleBlue, Droll, Dysmorodrepanis, E2eamon, Eclecticology, Edebraux, Eeekster, Ellmist, EntChickie, Epbr123, Epipelagic, Eric Forste, Escape Orbit, Excirial, Extransit, Fenevad, Fivedollarartist, Flewellyn, Frankie0607, Gadfium, Gaius Cornelius, GcSwRhIc, Gigemag76, Gilgamesh, Glen, Glenn, Globalsolidarity, Gmaxwell, Gouerouz, Gprince007, Graham87, Greg Howerton, Gregkaye, Gspbeetle, Gtg204y, Gurchzilla, Hadal, Haoie, Hibernian, Hlucho, Hydrogen Iodide, Iancrose, Igodard, Invertzoo, Iridescent, IronChris, Isnow, IvanLanin, Ixfd64, J.delanoy, JForget, Jaerik, Jebus989, Jeffrey Smith, Jeltz, Jj137, Joe Schmedley, Josh Grosse, Jsmaye, Juansidious, Jwmartin1, Karen Johnson, Kartano, Kasmeister, Kidd Loris, KnowledgeOfSelf, Kozuch, Kr0n05931, Kyleworj, La Pianista,

Lafuerzasindical, Lahiru k, Legion fi, LeoNomis, LibLord, Lights, Ligulem-s, Lugia2453, Lycaon, MER-C, Mac Dreamstate, Magister Mathematicae, Mani1, Marc Venot, Materialscientist, MauriceJFox3, Mav, McSly, Menchi, Mgiganteus1, Michael Hardy, MichaelAWilson, Michaelzeng7, Mike 7, Moebuggy, Mona, Moogle10000, Mouse Nightshirt, Munozmario01, Muriel Gottrop, Mysid, Mythsearcher, Neelix, Nehrams2020, Nepenthes, Nick Brakey, Nivix, Nneonneo, Noctibus, Nono64, Nubiatech, Nv8200p, Oanh, Od Mishehu, Omgbleh, Onco p53, Optigan13, Orphic, OverlordQ, Oxforduniv, Oxymoron83, Paxsimius, PenguinJockey, Peter coxhead, PeterNesbitt, Pfranson, Phle96, Piisamson, Pinethicket, Pingveno, Pmcalduff, Pmokeefe, Pne, Possum, Postoak, Proofreader77, Proteins, Quatermass, QuiteUnusual, Qwertzy2, RA0808, Radical Mallard, RainbowOfLight, Rainiermrc, RandomP, Raul654, Razorflame, Reach Out to the Truth, Rearete, Repairpages, RexNL, Rhadams, Rhodydog6, Rjwilmsi, Rocastelo, Romiet3625, Ronhjones, Roux, Rror, Rsocol, SMC, SUL, Sanfranman59, Saperaud, Scapermoya, Sceptre, SchrodingersKitten, Scottalter, Seb az86556, Seddon, Seglea, Shanes, Shoeofdeath, Slavy, SmartGuy Old, Smartse, Smith609, Smitty, Smooth O, Snowmanradio, Snoyes, Sobirà, SouthernNights, Stan Shebs, Stemonitis, Suffusion of Yellow, Summer Song, Swaggiesami, Taxman, Tcncv, Tedclaymore, Thatguyflint, The Thing That Should Not Be, Tide rolls, Tigershrike, Tim Q. Wells, Tiptoety, Tjunier, Tommy2010, Trevor MacInnis, Unyoyega, UtherSRG, Vainamainien, Vanished user wijr2nqfoijwvlks4tu9wcjoqiwf, Vietbio, WatermelonPotion, Wayne Slam, Weaseloid, Wilson44691, Wimt, Worldtraveller, Xiaphias, Yath, Yerpo, Yeshhh, Zfr, \\\\\\pwarg\\\\\\, 606 anonymous edits

Malacostraca *Source*: http://en.wikipedia.org/w/index.php?title=Malacostraca *Contributors*: Abyssal, Aledubr, Andycjp, Attilios, Aymankamelwiki, Azhyd, Beyond My Ken, Bgold, Carcinologist, Closedmouth, Craig Pemberton, Crustaceanguy, DabMachine, Dawynn, Delcrumbo, Dj Capricorn, Dougofborg, Fabien NGUYEN, Gdr, Glenn, GrahamBould, Gunnar Mikalsen Kvifte, HennessyC, J. Spencer, Kummi, Lady Tenar, LilHelpa, Locomiguel, LordJumper, Lupo, Lycaon, Mahlum, Makki98, Matticus78, MelbourneStar, Muriel Gottrop, Nm1234, Nono64, Poorleno, Pseudofusulina, Qwertzy2, RedWolf, Rhrad, Saperaud, Scottalter, Seglea, Squidonius, Squids and Chips, Stemonitis, Tau'olunga, Uncle G, UtherSRG, V111P, Vojtech.dostal, Voyevoda, WereSpielChequers, Xoristzatziki, לובי לב, 49 anonymous edits

Decapoda *Source*: http://en.wikipedia.org/w/index.php?title=Decapoda *Contributors*: 2001:db8, A-giau, A.shteiman, Ahoerstemeier, Al E., Arcadian, Arnejohs, Art LaPella, Azhyd, Bellenion, Betacommand, Brrrtje, Chank12, Crustaceanguy, Dante Alighieri, Dj Capricorn, Epipelagic, Fryed-peach, Gary12345, Gdr, Gigemag76, Glenn, Gouerouz, Guessing Game, Guqin, Hellbus, IdLoveOne, IronChris, Janderk, JaviPrieto, Jimp, JoJan, Jwmartin1, Kku, Lady Tenar, Ligulem-s, Lycaon, Mani1, Mgiganteus1, Morwen, Mumbo jiggy, Pgan002, PoccilScript, Postdlf, Qwertzy2, Rui Silva, Seb az86556, Seglea, Splodgeness, Squidonius, Stan Shebs, Stemonitis, Supersaiyanplough, T34, Tau'olunga, Template namespace initialisation script, Ufim, Vaqo, Vietbio, Voyevoda, Yerpo, ZhiRandom, 41 anonymous edits

Image Sources, Licenses and Contributors

file:Liocarcinus_vernalis.jpg *Source*: http://en.wikipedia.org/w/index.php?title=File:Liocarcinus_vernalis.jpg *License*: unknown *Contributors*: Amada44, Berrucomons, Biopics, Deadstar, JuTa, Lycaon, MrPanyGoff, Stemonitis, Via null, 1 anonymous edits

File:Gecarcinus quadratus (Nosara).jpg *Source*: http://en.wikipedia.org/w/index.php?title=File:Gecarcinus_quadratus_(Nosara).jpg *License*: unknown *Contributors*: (Bhny)

File:Pachygrapsus marmoratus male female.png *Source*: http://en.wikipedia.org/w/index.php?title=File:Pachygrapsus_marmoratus_male_female.png *License*: unknown *Contributors*: User:George Chernilevsky, User:Stemonitis

File:Xantho poressa 2009 G1.jpg *Source*: http://en.wikipedia.org/w/index.php?title=File:Xantho_poressa_2009_G1.jpg *License*: unknown *Contributors*: User:George Chernilevsky

File:Carpilius convexus is consuming Heterocentrotus trigonarius in Hawaii.jpg *Source*: http://en.wikipedia.org/w/index.php?title=File:Carpilius_convexus_is_consuming_Heterocentrotus_trigonarius_in_Hawaii.jpg *License*: unknown *Contributors*: Brocken Inaglory

File:Sorting Crabs Ffionphort.jpg *Source*: http://en.wikipedia.org/w/index.php?title=File:Sorting_Crabs_Ffionphort.jpg *License*: unknown *Contributors*: User:Velela

File:Crabmasala.jpg *Source*: http://en.wikipedia.org/w/index.php?title=File:Crabmasala.jpg *License*: unknown *Contributors*: Charles Haynes

Image:Dromia personata.jpg *Source*: http://en.wikipedia.org/w/index.php?title=File:Dromia_personata.jpg *License*: unknown *Contributors*: Amada44, Biopics, Lycaon, Via null

Image:Ranina ranina.jpg *Source*: http://en.wikipedia.org/w/index.php?title=File:Ranina_ranina.jpg *License*: unknown *Contributors*: User:Kzhr

Image:Corystes cassivelaunus.jpg *Source*: http://en.wikipedia.org/w/index.php?title=File:Corystes_cassivelaunus.jpg *License*: unknown *Contributors*: AFBorchert, Achird, Amada44, Biopics, Jon Harald Søby, Lycaon, Maashatra11, NeverDoING, Nikkirikki, Nilfanion, Sanomi, Via null, 6 anonymous edits

Image:Ocypode quadrata (Martinique).jpg *Source*: http://en.wikipedia.org/w/index.php?title=File:Ocypode_quadrata_(Martinique).jpg *License*: unknown *Contributors*: Patrick Verdier Original uploader was Alkapote at fr.wikipedia

file:Atergatis subdentatus.jpg *Source*: http://en.wikipedia.org/w/index.php?title=File:Atergatis_subdentatus.jpg *License*: unknown *Contributors*: Dr. Dwayne Meadows, NOAA/NMFS/OPR

file:Animal diversity.png *Source*: http://en.wikipedia.org/w/index.php?title=File:Animal_diversity.png *License*: unknown *Contributors*: User:Anilocra, User:Bkmiles, User:Kevincollins123, User:Lviatour, User:Medeis, User:Nhobgood, User:Opoterser, User:Panda3, user:Australianplankton

File:Mitosis-fluorescent.jpg *Source*: http://en.wikipedia.org/w/index.php?title=File:Mitosis-fluorescent.jpg *License*: unknown *Contributors*: Deadstar, Glenn, Hikke, Iori951, Kam Solusar, Linél, Monkeybait, Mortadelo2005, Opabinia regalis, Pieter Kuiper, Sanbec, Steveprutz, Túrelio, 10 anonymous edits

File:Dunkleosteus BW.jpg *Source*: http://en.wikipedia.org/w/index.php?title=File:Dunkleosteus_BW.jpg *License*: unknown *Contributors*: Nobu Tamura (http://spinops.blogspot.com)

File:Vernanimalcula.jpg *Source*: http://en.wikipedia.org/w/index.php?title=File:Vernanimalcula.jpg *License*: unknown *Contributors*: Moe Epsilon, Ricky81682, Smith609

File:AnimalsRelativeNumbers.png *Source*: http://en.wikipedia.org/w/index.php?title=File:AnimalsRelativeNumbers.png *License*: unknown *Contributors*: Nick Beeson (Nwbeeson at en.wikipedia)

File:Elephant-ear-sponge.jpg *Source*: http://en.wikipedia.org/w/index.php?title=File:Elephant-ear-sponge.jpg *License*: unknown *Contributors*: Elapied, Linél, Maksim

File:Superbfairywrenscropped.jpeg *Source*: http://en.wikipedia.org/w/index.php?title=File:Superbfairywrenscropped.jpeg *License*: unknown *Contributors*: Benjamint444, CommonsDelinker, Pengo, Smihael, Tony Wills, Wst, 1 anonymous edits

File:Sympetrum flaveolum - side (aka).jpg *Source*: http://en.wikipedia.org/w/index.php?title=File:Sympetrum_flaveolum_-_side_(aka).jpg *License*: unknown *Contributors*: user:Aka

File:Bedford's Flatworm.jpg *Source*: http://en.wikipedia.org/w/index.php?title=File:Bedford's_Flatworm.jpg *License*: unknown *Contributors*: Jan Derk

File:Grapevinesnail 01.jpg *Source*: http://en.wikipedia.org/w/index.php?title=File:Grapevinesnail_01.jpg *License*: unknown *Contributors*: Jürgen Schoner

File:Carolus Linnaeus (cleaned up version).jpg *Source*: http://en.wikipedia.org/w/index.php?title=File:Carolus_Linnaeus_(cleaned_up_version).jpg *License*: unknown *Contributors*: Original painting by Alexander Roslin. Digitally improved by Greg L.

file:Arthropoda.jpg *Source*: http://en.wikipedia.org/w/index.php?title=File:Arthropoda.jpg *License*: unknown *Contributors*: Amada44, Bugboy52.40, CommonWikiMan, Hekerui, Xvazquez, 3 anonymous edits

Image:HemimerusHanseni.jpg *Source*: http://en.wikipedia.org/w/index.php?title=File:HemimerusHanseni.jpg *License*: unknown *Contributors*: Maxwell Lefroy

Image:Biramous cross section 01.png *Source*: http://en.wikipedia.org/w/index.php?title=File:Biramous_cross_section_01.png *License*: unknown *Contributors*: User:Philcha

Image:Arthropod_head_problem_02.png *Source*: http://en.wikipedia.org/w/index.php?title=File:Arthropod_head_problem_02.png *License*: unknown *Contributors*: User:Philcha

Image:Arthropod cuticle 01.png *Source*: http://en.wikipedia.org/w/index.php?title=File:Arthropod_cuticle_01.png *License*: unknown *Contributors*: User:Philcha

File:Molt4.jpg *Source*: http://en.wikipedia.org/w/index.php?title=File:Molt4.jpg *License*: unknown *Contributors*: NOAA

Image:Arthropod body struct 01.png *Source*: http://en.wikipedia.org/w/index.php?title=File:Arthropod_body_struct_01.png *License*: unknown *Contributors*: Philcha (talk) Original uploader was Philcha at en.wikipedia

File:Wasp ocelli.JPG *Source*: http://en.wikipedia.org/w/index.php?title=File:Wasp_ocelli.JPG *License*: unknown *Contributors*: Genesis12, Patrick, Thebugman123, 2 anonymous edits

File:scorpionwithyoung.JPG *Source*: http://en.wikipedia.org/w/index.php?title=File:Scorpionwithyoung.JPG *License*: unknown *Contributors*: Original uploader was Fusion121 at en.wikipedia

File:Shrimp nauplius.jpg *Source*: http://en.wikipedia.org/w/index.php?title=File:Shrimp_nauplius.jpg *License*: unknown *Contributors*: Spedona, Stemonitis,

File:Marrella.jpg *Source*: http://en.wikipedia.org/w/index.php?title=File:Marrella.jpg *License*: unknown *Contributors*: Ghedoghedo, Haplochromis, Origamiemensch

File:Onycophora.jpg *Source*: http://en.wikipedia.org/w/index.php?title=File:Onycophora.jpg *License*: unknown *Contributors*: Bruno Vellutini from São Paulo/São Sebastião, Brasil

File:Insect food stall.JPG *Source*: http://en.wikipedia.org/w/index.php?title=File:Insect_food_stall.JPG *License*: unknown *Contributors*: User:Takoradee

file:Abludomelita obtusata.jpg *Source*: http://en.wikipedia.org/w/index.php?title=File:Abludomelita_obtusata.jpg *License*: unknown *Contributors*: Amada44, Berrucomons, Biopics, Lycaon, Slaunger, Via null

File:Crab from Long Island.jpg *Source*: http://en.wikipedia.org/w/index.php?title=File:Crab_from_Long_Island.jpg *License*: unknown *Contributors*: User:Invertzoo

File:Krillanatomykils.jpg *Source*: http://en.wikipedia.org/w/index.php?title=File:Krillanatomykils.jpg *License*: unknown *Contributors*: Kersti Nebelsiek, Mithril, Nature Loader, Perhelion, Saperaud, Smartse, Stemonitis, Uwe kils, 1 anonymous edits

File:Potamon fluviatile9.jpg *Source*: http://en.wikipedia.org/w/index.php?title=File:Potamon_fluviatile9.jpg *License*: unknown *Contributors*: User:Esculapio

File:Homarus gammarus zoea.jpg *Source*: http://en.wikipedia.org/w/index.php?title=File:Homarus_gammarus_zoea.jpg *License*: unknown *Contributors*: Amada44, Biopics, Lycaon, NeverDoING, TheF, Via null

File:Haeckel Copepoda.jpg *Source*: http://en.wikipedia.org/w/index.php?title=File:Haeckel_Copepoda.jpg *License*: unknown *Contributors*: Biopics, CommonsDelinker, Dysmorodrepanis, Lycaon, Pengo, Ragesoss, 1 anonymous edits

File:Haeckel Decapoda.jpg *Source*: http://en.wikipedia.org/w/index.php?title=File:Haeckel_Decapoda.jpg *License*: unknown *Contributors*: Biopics, Dysmorodrepanis, Pengo, Ragesoss, 1 anonymous edits

File:Daphnia pulex.png *Source*: http://en.wikipedia.org/w/index.php?title=File:Daphnia_pulex.png *License*: unknown *Contributors*: (Photo: Paul Hebert)

File:Speleonectes tanumekes unlabeled-rotated.png *Source*: http://en.wikipedia.org/w/index.php?title=File:Speleonectes_tanumekes_unlabeled-rotated.png *License*: unknown *Contributors*: User:Peter coxhead

File:Chthamalus stellatus.jpg *Source*: http://en.wikipedia.org/w/index.php?title=File:Chthamalus_stellatus.jpg *License*: unknown *Contributors*: User:MichaelMaggs

File:Ostracod.JPG *Source*: http://en.wikipedia.org/w/index.php?title=File:Ostracod.JPG *License*: unknown *Contributors*: Original uploader was Anna33 at en.wikipedia

File:Gammarus roeselii.jpg *Source*: http://en.wikipedia.org/w/index.php?title=File:Gammarus_roeselii.jpg *License*: unknown *Contributors*: User:snek01

File:Eryma mandelslohi (Krebs) - Oberer Brauner Jura - Bissingen unter Teck.jpg *Source*: http://en.wikipedia.org/w/index.php?title=File:Eryma_mandelslohi_(Krebs)_-_Oberer_Brauner_Jura_-_Bissingen_unter_Teck.jpg *License*: unknown *Contributors*: user:Ra'ike

File:Shrimps at market in Valencia.jpg *Source*: http://en.wikipedia.org/w/index.php?title=File:Shrimps_at_market_in_Valencia.jpg *License*: unknown *Contributors*: 1029man, FlickreviewR, Juiced lemon, Kri, Linél, Luis Fernández García, Melimama, Salix, TwoWings

Image:wikisource-logo.svg *Source*: http://en.wikipedia.org/w/index.php?title=File:Wikisource-logo.svg *License*: unknown *Contributors*: Guillom, INeverCry, Jarekt, MichaelMaggs, NielsF, Rei-artur, Rocket000

file:Crustacea.jpg *Source*: http://en.wikipedia.org/w/index.php?title=File:Crustacea.jpg *License*: unknown *Contributors*: User:Biopics

File:Sally lightfoot crab.jpg *Source*: http://en.wikipedia.org/w/index.php?title=File:Sally_lightfoot_crab.jpg *License*: unknown *Contributors*: User:Barfbagger

File:Nebalia bipes.jpg *Source*: http://en.wikipedia.org/w/index.php?title=File:Nebalia_bipes.jpg *License*: unknown *Contributors*: Amada44, Berrucomons, Biopics, Lycaon, NeverDoING, Sanomi, Via null, 5 anonymous edits

File:Odontodactylus scyllarus3.jpg *Source*: http://en.wikipedia.org/w/index.php?title=File:Odontodactylus_scyllarus3.jpg *License*: unknown *Contributors*: Jens Petersen

File:Porcellio scaber and Oniscus asellus - Zalné20070205.jpg *Source*: http://en.wikipedia.org/w/index.php?title=File:Porcellio_scaber_and_Oniscus_asellus_-_Zalné20070205.jpg *License*: unknown *Contributors*: User:Pudding4brains

File:Cancer pagurus.jpg *Source*: http://en.wikipedia.org/w/index.php?title=File:Cancer_pagurus.jpg *License*: unknown *Contributors*: Achird, Amada44, Biopics, Gryffindor, Lycaon, Stemonitis, Via null

file:Haeckel Decapoda.jpg *Source*: http://en.wikipedia.org/w/index.php?title=File:Haeckel_Decapoda.jpg *License*: unknown *Contributors*: Biopics, Dysmorodrepanis, Pengo, Ragesoss, 1 anonymous edits

File:Penaeus vannamei 01.jpg *Source*: http://en.wikipedia.org/w/index.php?title=File:Penaeus_vannamei_01.jpg *License*: unknown *Contributors*: Linél, Lycaon, WeFt

File:Heterocarpus ensifer.jpg *Source*: http://en.wikipedia.org/w/index.php?title=File:Heterocarpus_ensifer.jpg *License*: unknown *Contributors*: NOAA

File:Austropotamobius pallipes.jpg *Source*: http://en.wikipedia.org/w/index.php?title=File:Austropotamobius_pallipes.jpg *License*: unknown *Contributors*: David Gerke

File:Upogebia deltaura.jpg *Source*: http://en.wikipedia.org/w/index.php?title=File:Upogebia_deltaura.jpg *License*: unknown *Contributors*: Amada44, Berrucomons, Biopics, Linél, Lycaon, Nevit

File:Lobster 300.jpg *Source*: http://en.wikipedia.org/w/index.php?title=File:Lobster_300.jpg *License*: unknown *Contributors*: JoJan, Linél, Lucero del Alba, 2 anonymous edits

File:Polycheles sculptus.jpg *Source*: http://en.wikipedia.org/w/index.php?title=File:Polycheles_sculptus.jpg *License*: unknown *Contributors*: SEFSC Pascagoula Laboratory; Collection of Brandi Noble, NOAA/NMFS/SEFSC

File:CoenobitaVariabilis2.jpg *Source*: http://en.wikipedia.org/w/index.php?title=File:CoenobitaVariabilis2.jpg *License*: unknown *Contributors*: Vanessa Pike-Russell

File:Blue crab on market in Piraeus - Callinectes sapidus Rathbun 20020819-317.jpg *Source*: http://en.wikipedia.org/w/index.php?title=File:Blue_crab_on_market_in_Piraeus_-_Callinectes_sapidus_Rathbun_20020819-317.jpg *License*: unknown *Contributors*: User:Wpopp

MIX
Papier aus verantwortungsvollen Quellen
Paper from responsible sources
FSC® C105338

Printed by Books on Demand GmbH, Norderstedt / Germany